AF725686

DISEASE AND THE COMPENSATION DEBATE

DISEASE
AND THE
COMPENSATION DEBATE

JANE STAPLETON

CLARENDON PRESS · OXFORD

This book has been printed digitally and produced in a standard specification in order to ensure its continuing availability

OXFORD
UNIVERSITY PRESS

Great Clarendon Street, Oxford OX2 6DP

Oxford University Press is a department of the University of Oxford.
It furthers the University's objective of excellence in research, scholarship,
and education by publishing worldwide in

Oxford New York

Auckland Bangkok Buenos Aires Cape Town Chennai
Dar es Salaam Delhi Hong Kong Istanbul Karachi Kolkata
Kuala Lumpur Madrid Melbourne Mexico City Mumbai Nairobi
São Paulo Shanghai Singapore Taipei Tokyo Toronto

with an associated company in Berlin

Oxford is a registered trade mark of Oxford University Press
in the UK and in certain other countries

Published in the United States
by Oxford University Press Inc., New York

© Jane Stapleton 1986

The moral rights of the author have been asserted
Database right Oxford University Press (maker)

Reprinted 2002

All rights reserved. No part of this publication may be reproduced,
stored in a retrieval system, or transmitted, in any form or by any means,
without the prior permission in writing of Oxford University Press,
or as expressly permitted by law, or under terms agreed with the appropriate
reprographics rights organization. Enquiries concerning reproduction
outside the scope of the above should be sent to the Rights Department,
Oxford University Press, at the address above

You must not circulate this book in any other binding or cover
and you must impose this same condition on any acquirer

ISBN 0-19-825552-7

Preface

THIS book is a revised version of a thesis submitted for the degree of D. Phil. in the University of Oxford. I would particularly like to thank my supervisor, Professor P. S. Atiyah, for his generosity with time and advice.

JANE STAPLETON

14th March, 1986

List of Tables

List of Figures

Contents

Abbreviations

Atiyah, *ACL*	P. S. Atiyah, *Accidents, Compensation and the Law*, 3rd ed. (London, Weidenfeld & Nicholson, 1980).
Pearson	*Royal Commission on Civil Liability and Compensation for Personal Injury* (chm. Lord Pearson) (March 1978) Cmnd. 7054 (Vol. 1 unless otherwise specified).
Woodhouse Australia	*Compensation and Rehabilitation in Australia: Report of the Committee of Inquiry* (Canberra, AGPS, 1974).
Woodhouse New Zealand	*Report of the Royal Commission of Inquiry, Compensation for Personal Injury* (chm. Woodhouse J.) (Wellington, NZGP, 1967).

Table of Cases

Table of Decisions

Table of Statutes and Statutory Instruments

1

Man-Made Disease

1.1 *Introduction*

In the past twenty years the law concerning compensation for unintentionally inflicted personal injuries, especially the remedy in tort, has attracted considerable interest and debate among lawyers and social reformers. The principal aim of this book is to argue that serious distortions underlie this debate because of its tendency to focus on accidents and to neglect issues raised by other types of personal injury which cannot be unequivocally pin-pointed to a certain time or place but, because they arise from a man-made source, could potentially give rise to a claim in tort. These actual or apparently non-traumatic injuries—which I will loosely term 'man-made disease'—are not unusual. They include, for example: occupational injuries such as dust diseases, repetitive movement injury, deafness, and dermatitis; product-caused injuries such as chronic adverse drug reactions and tobacco-caused diseases; and environmental injuries such as diseases caused by fertilizers and man-made radiation. Indeed, many of the *causes célèbres* which have recently fuelled interest in the reform or replacement of tort law in the personal injuries field have involved hazards which fall into this broad class of injury: asbestos, thalidomide, diethylstilbestrol (DES), and Agent Orange.

Despite this, a survey of the major contributions to the modern literature on compensation reveals an almost exclusive concern with accidents. In New Zealand, for example, the early concern which surfaced in the Report of the Committee on Absolute Liability (1963) and later in the Report of the Royal Commission (1967), set up under Woodhouse J. to look at industrial compensation, was the desirability of enacting a compensation scheme to cover the whole class (but only the class) of accidents. As a consequence the resultant New Zealand scheme—the principal operating model of a comprehensive non-tort compensation scheme—is primarily an accident scheme. In Australia the Whitlam Labor Government (1972–5) was committed to a similar accident scheme, but charged the Committee of Inquiry handling its formulation—also chaired by Woodhouse J.—with the added task of looking into the possibility of covering all disability. Yet, while the Committee's Report acknowledged the desirability of this and designed a scheme to accommodate it, when priorities were being considered accident coverage was again given primary importance in the proposals. Even at the close of the 1970s, Woodhouse himself still appeared to see reform as principally concerned with accident compensation.[1] Indeed, the idea of a compensation scheme covering all disability, including man-made and 'naturally occurring' disease, was so unsuccessful in capturing the imagination of reformers that under the current proposals of the Hawke Labor Government in Australia non-occupational disease victims no longer feature on the reform agenda at all. What is being proposed is a step by step reform culminating in an accident scheme like that in New Zealand—there is no longer even an avowed long-term commitment to include other disabled groups.

In Britain and North America governments have been slower than the Australasians to investigate reform in the personal injuries area, but academics and other lawyers have been vocal since the early sixties in their criticism of current compensation systems. Initially these studies of the tort remedy concentrated on road accidents, and even when they later gave rise to more general studies they were again, and still are, principally concerned with accidents and traumatically caused injuries.[2] Perhaps the best illustration of the attraction this accident focus has exerted on reformers in the compensation debate is the way in which the issues raised by the dramatic thalidomide injuries (to children born between 1959 and 1962 in Britain) were rarely addressed by commentators in the 1960s, and were often overlooked by the very government inquiry bodies on personal injury compensation, such as the Pearson Commission, which had been set up in part as a response to that particular drug disaster (see 5.5 and 5.8). Finally,

[1] A. Woodhouse, *A Challenge to the Law: Personal Injuries Compensation* (Calgary, Burroughs & Co., 1979), 11–12, 26.

[2] See e.g. J. Jolowicz, 'Liability for Accidents' [1968] *Camb. L. J.* 50; T. Ison, *The Forensic Lottery* (London, Staples Press, 1967); Atiyah, *ACL* and early road accident studies cited therein. Cf. T. Ison, *Accident Compensation* (London, Croom Helm, 1980), ch. 2, and Harris *et al.*, *Compensation and Support for Illness and Injury* (Oxford, Clarendon Press, 1984).

in the United States where there is much less enthusiasm for sweeping government intervention in the compensation of personal injuries, the few analyses and reform proposals made by legal commentators display a comparable concentration on accidents. An additional interesting feature here is that the more popular and vigorous debate in that country on the deterrent aspect of legal rules in the area, is similarly preoccupied with traumatic injuries and this has had just as important a distorting effect on the analysis as the accident preference has had on the compensation debate (see Chapter 6).

This bias towards accidents in readily explained for, although the law of torts does not formally sanction preferential treatment of accident victims, its conceptual machinery dramatically favours this class in practice. As a consequence *effective* tort liability is principally confined to cases of traumatic injury. Ironically, this has led critics of the tort system to concentrate on accidents and to neglect some of the most damaging arguments at their disposal by overlooking the greater barriers the system presents to victims of negligence which has happened to result in non-traumatic rather than traumatic injury. Similarly, analysts and defenders of the tort system who focus on accident cases fail to identify cases of non-traumatic injury as the catalysts they have been for revolutionary reforms and innovations which provide testament to the flexibility of the existing system. The result of this neglect is to distort both the debate on the adequacy of the law concerning compensation for personal injuries and the reform proposals to which it gives rise.

This book, then, sets out to explore issues raised by cases of non-traumatic injury. It examines the conceptual and practical problems faced by victims of man-made disease in recovering damages in tort, and how these difficulties are not satisfactorily resolved by reforms such as the introduction of stricter liability or non-tort schemes limited on the basis of the cause of injury. Only a non-tort scheme based on disability rather than cause would seem to overcome these difficulties. This ultimately raises the question of how the reform agenda would be transformed by an awareness of the issues raised by non-traumatic injury, not only within the debate over the adequacy of tort law in its compensatory and deterrence functions, but also in the broader debate on public schemes to provide for the disabled—their possible form, rationale, and general relation to systems of support for victims of other forms of misfortune.

1.2 *Disablement: Some Definitions and Statistics*

At the outset it is useful and important to classify cases of disablement on the basis of how they were caused. Broadly speaking, disablement can result from injury from one of three different types of cause (see Figure 1.1). First, there are injuries which are due to accidents (class 1). Secondly, there are injuries traceable to a non-traumatic but man-made source which might, therefore, in principle give rise to a claim in tort (class 2). It is with this group that most of the early discussion will be concerned. Thirdly, there are cases of disablement which cannot be described in either of these ways and which are grouped together under the term of disability due to 'natural causes' (class 3), although it is important to emphasize that this class includes many cases where there is in fact a man-made cause not yet proven or suspected.

Of course the demarcation lines between these classes are artificial from a medical point of view but for various reasons they provide workable classifications for legal analysis. Thus the line

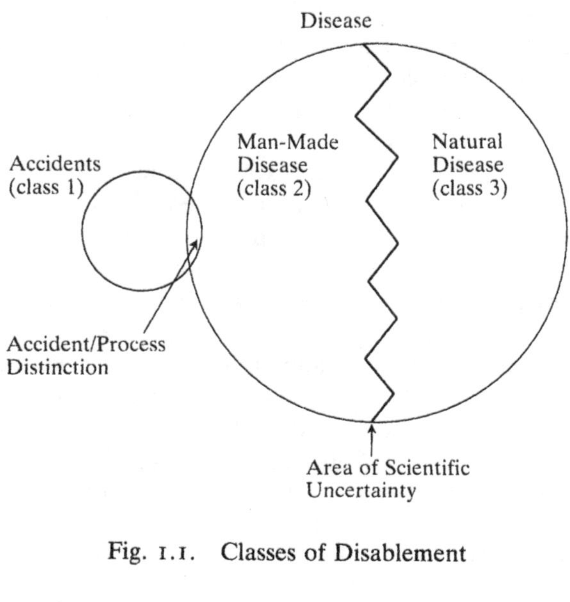

Fig. 1.1. Classes of Disablement

between class 1 and the other classes is basically that an 'accident' occurs at an isolated and identifiable time and place whereas non-traumatic injury does not. On this basis, for example, cases of nervous shock and acute food poisoning are classifiable as accidents while diseases of infection, where the time and place of infection are not identifiable, are best seen as cases of non-traumatic injury. This would also be the case with latent adverse drug reactions, although most other examples of medical misadventure would be classified as accidents because the damage is referable to a known point in time, for example post-operative complications. This distinction between the class of accidents and other causes of disability does throw up the accident/process anomalies familiar to those concerned with industrial compensation (see 3.6), but it is a crudely workable distinction since relatively few cases fall within this penumbra of definitional uncertainty. This is not the case with the line between classes 2 and 3. Here there is the intrinsic threshold problem of defining what factors are 'man-made'. For example, are foetal abnormalities traceable to a parent's pre-natal regime (in diet, exercise, smoking habits, etc.) man-made? Are deaths due to hypothermia, or to the inadequacies of medical facilities available in the individual case, to be classed as man-made alongside lung cancer from smoking or asbestos dust inhalation? Even if these definitional questions could be answered, there would be a great number of cases which are impossible to attribute to either a man-made or 'natural' source. In Chapter 7 it will be seen that this unworkability of the distinction between classes 2 and 3 has important ramifications for social policy, but at this stage it is sufficient to note the usefulness of a collective term for disability caused other than by accident. This will be called 'disease' made up of classes 2 ('man-made disease') and 3 ('natural disease'). Of course these classes include cases of muscular conditions, congenital abnormalities, and the like which are not referred to in common usage as 'disease' so the terminology here is slightly unconventional. It is, however, a convenient shorthand which does not involve any significant misdescription.

Given these three classes of disabled, can some idea be obtained of the relative number of cases involved? There are well over three million people with a physical or mental handicap living in private households in Britain and, of these, by far the least numerous class of disabled is class 1, those due to accidents, which accounts for only about 10 per cent of the disabled (and 3 per cent of deaths).[3] This means that any compensation system covering all disability would be overwhelmingly concerned with disease cases rather than accident victims and this has profound implications for the form and rationale of such a scheme (see 7.5). But since the discussion will initially be confined to questions concerning the adequacy of the tort system, it is important to get some idea of the volume of disability cases that that system deals with in theory and in practice. In theory, disability attributable to the fault of (or, in a few cases, simply caused by) a human agent may give rise to an action for damages in tort. The potential catchment of tort, therefore, extends beyond the 10 per cent of cases of disability which are caused by accident to include class 2 cases such as diseases due to industrial, product, or environmental health hazards. It is impossible to say just how large is this latter class of actionable man-made disease. There has simply been no concerted attempt so far to collect the requisite statistics. Figures cited by reform bodies such as the Pearson Commission are unhelpful because of the ambiguous use of the term 'injury'. For example, while clearly being used in many parts of that Report synonymously with 'traumatically caused injury', and thereby to limit the inquiry to accidents, many instances of non-traumatic sources of disability, such as chronic adverse drug reactions, are included in the discussion. This throws doubt on the basis of the statistical estimates of the number of 'injuries' in total and from various sources. The approach, itself an indication of the neglect typically accorded man-made disease, leaves unclear both the size of this class of injury, and, consequently, the importance of reformers overlooking the issues of man-made disease in the compensation debate and reform proposals. (In this book 'injury' will include disability from both traumatic and non-traumatic causes.) Even when a distinction between accidents and disease is made—as, for example, when a compensation scheme is expressly designed to cover only the former—a statistical distinction within the excluded classes of disease between those of man-made and 'natural' origin is not made, so that it is impossible to gauge how many cases of

[3] Department of Health and Social Security, *Industrial Injuries Compensation: A Discussion Document* (London, DHSS, 1980), 9 (1968 figures); Pearson, paras. 41, 1514–20; vol. 2, para. 35.

man-made disease are thereby being excluded from the scheme. Nor are the considerable statistics available on *avoidable* disease useful because they include factors which relate to personal life-style such as failure to exercise and not to the potentially tortious conduct of another. There are, however, a few isolated reliable statistics which indicate, if not the absolute size of the class of potentially actionable man-made disease, at least the fact that it is not a numerically insignificant group of the disabled. For example:[4]

(i) Each year about 12,000 reports of adverse drug reactions are made to the Committee on the Safety of Medicines (CSM) via the yellow card system and even that number is generally acknowledged to be a gross underestimate of the true figure. A dramatic example of this in the early years of the system has been given by Inman, a former Senior Medical Officer with the CSM. Inman found that although there had been 3,500 excess deaths of asthmatics in England and Wales in the period 1961–7 which had probably been caused by the use of pressurized aerosols, up until 1967 only six yellow card reports had been made to the CSM.

(ii) Non-medicinal products also cause disease. For example, a recent Australian government study found that addictive products such as tobacco and alcohol are implicated in 20 per cent of all deaths in that country.

(iii) While there is no reliable estimate of the total number of

[4] (i) Committee on the Safety of Medicines, *Annual Report 1984* (London, CSM, 1985), 9; the figure includes some acute (traumatic) reactions. W. Inman and A. Adelstein, 'Rise and Fall of Asthma Mortality in England and Wales in Relation to the Use of Pressurized Aerosols' [1969] 2 *Lancet*, 279; W. Inman, 'Role of Drug-Reaction Monitoring in the Investigation of Thrombosis and the Pill' (1970) 26 *British Medical Bull.* 248. (ii) National Information Service on Drug Abuse, *Technical Information Bulletin* (No. 69, Oct. 1982) (Canberra, AGPS, 1983), 3. (iii) M. Crick and G. Linsley, *An Assessment of the Radiological Impact of the Windscale Reactor Fire, October 1957* (National Radiological Protection Board, NRPB-R135, Nov. 1982), 23. (iv) R. Coles, 'Medico-Legal Aspects of Noise Hazards to Hearing' (1975) 43 *Medico-Legal J.* 3. (v) Anon., 'Compensating Victims of Occupational Disease' (1980) 93 *Harv. LR.* 916. R. Doll and R. Peto, *The Causes of Cancer* (Oxford, OUP, 1981), 1245 (US estimate); Health and Safety Executive, *Technical Report No. 1: Carcinogens in the Workplace* (HMSO, 1982), 1–2 (that 1% of cancer deaths corresponds to a figure of 1,400 deaths). (vi) Health and Safety Executive, *Health and Safety Statistics 1981–1982* (London, HMSO, 1985); R. Peto, letter to *New Statesman*, 10 Sept. 1982, 18. Imperial Cancer Research Fund, *Cancer Epidemiology and Clinical Trials Unit* (London, ICRF, June 1983), 1, 3 (UK figures); Doll and Peto, *The Causes of Cancer*, 1224–5 (US figures).

cancer cases due to radiation leaks from nuclear installations, one government estimate states that a single leak from the Windscale (Sellafield) plant in 1957 has probably resulted in an extra 250 cases of thyroid cancer in Britain.

(iv) Two million people in Britain are affected by noise deafness.

(v) The higher exposure to health hazards in the workplace and better record keeping have produced a more consistent stream of statistics of disease from this man-made source. Though subject to considerable qualification, such statistics suggest that occupational disease accounts for 100,000 deaths and 390,000 new cases of disability each year in the United States (corresponding estimates for Britain are unavailable). Indeed, even on the most conservative estimates, workplace exposures account for 4 per cent of all cancer deaths, that is, for approximately 5,600 deaths in Britain and 17,000 deaths in the US each year.

(vi) But perhaps the most dramatic of the available statistics concern just two products, asbestos and tobacco. One of the fatal asbestos-related cancers, mesothelioma, alone causes around 500 deaths a year in Britain. A leading British epidemiologist, Richard Peto of Oxford University, has estimated that the current overall death rate due to asbestos in Britain is about 1,000 a year, a figure which can be expected to rise to a peak of about 3,000, so that over the next thirty years 50,000 people in Britain will die from this one man-made cause. The Oxford team of epidemiologists have made a particular study of tobacco and found that it can reliably be said to cause 30 per cent of all cancer deaths in Britain and the United States, ten times as many as the next largest reliably known effect which is that of another man-made product, alcohol. When tobacco's contribution to the death rate from non-cancer causes is included, this leads to an estimate that tobacco causes about 100,000 of the 670,000 deaths in Britain per year.

When compared to the 3 per cent of deaths (21,000) due to accidents per year it is clear even from these patchy statistics that a significant class of disease is reliably attributable to man-made sources and provides a potential catchment for tort claims. If more comprehensive statistics were available, and if causal attribution both at a medical and policy level were more refined, the class of actionable man-made disease would, of course, be shown to be even larger.

1.3 *The Variety and Characteristics of Man-Made Disease*

Before broaching the problems of compensating man-made disease it is useful to have in mind the variety of situations in which these occur and the particular characteristics which distinguish these injuries from accident cases. Medical conditions produced by the products and activities of human beings range from cancers to deafness, from dermatitis to heart disease, and include areas of great scientific uncertainty such as genetic and foetal damage (mutagenicity and teratogenicity respectively). Sources of disability include chemicals; physical hazards such as pressure, temperature, noise, light, vibration, radiation, and unhealthy work patterns such as shiftwork and fast repetitive movements (the latter causing conditions like tenosynovitis); and biological hazards such as microbes in humidifier or air-conditioning systems (which appear to be the source of the dramatic Legionnaires' Disease) and infectious diseases (such as Acquired Immune Deficiency Syndrome: AIDS). Exposure to the hazard may take place in the workplace, through use of a product, or by environmental pollution. Well-known examples of the first class of exposure are industrial dusts such as asbestos, and industrial chemicals such as vinyl chloride monomer. The class of hazardous consumer products includes not only dangerous pharmaceuticals such as thalidomide, diethylstilbestrol (DES), Eraldin, and whooping cough vaccine but non-medicinal products such as tobacco, alcohol, cosmetics, food additives, detergents, tampons, intra-uterine devices, fire retardant chemicals in clothing, and building materials such as asbestos and formaldehyde insulation foam. Typically the product is hazardous to health as designed, so that the resultant disease is said to be attributable to a 'design defect', but there are also examples of man-made diseases being caused by manufacturing errors in products such as contamination of clothing, food, or cooking oil. Thirdly, man-made diseases can be contracted from environmental sources, including: pollution of the air, for example by lead in petrol or by fumes from insulation foam; and pollution of water, for example by industrial effluent being discharged directly into watercourses or leeching indirectly from toxic waste dumps and chemically treated farmland. The soil itself can be contaminated, for example by the accidental use of toxic waste as 'filler' for road repairs, and entire food-chains can become affected by chemical pollution as happened in Michigan in 1973, where the accidental feeding of toxic chemicals to dairy cattle resulted in 97 per cent of the state's human population being contaminated by the hazardous chemical.[5]

However, except in a very general sense it is not useful to categorize hazards as, say, 'industrial' or 'environmental' because of the many indirect ways in which exposure can occur. For example, a hazard like asbestos dust causes disease not only to those who work with it directly but, because it is a carcinogen with no known safe level of exposure, it can also cause disease to white-collar workers in the same plant, to residents living nearby, and to cohabitees of workers who come into contact with the dust from the workers' clothes.[6] Similarly, tobacco can cause disease not only in consumers but to those working or living with them, while cervical cancer in women may be linked with the job of her male partner.[7] The production and application of toxic agricultural chemicals not only affects those engaged in the process but residents and users of the affected and surrounding land. If the chemical is slow to degrade it may pollute ground water, soil, and vegetation, entering the food-chain and thereby affecting individuals with no obvious temporal or spatial connection with the hazard. The same is true of the dumping of hazardous industrial wastes. Later chapters will describe how these phenomena produce striking anomalies in non-tort compensation schemes limited to certain categories of injury, typically those of an industrial origin.

What, then, are the characteristics of man-made disease which prevent success in the tort system in all but a tiny fraction of claims? Essentially these concern, firstly, the time frame within which disability is contracted and then manifested, and secondly the complex causal origin of non-traumatic injury. An accident occurs at an isolated and identifiable point in time and space

[5] J. Egginton, *Bitter Harvest* (London, Secker & Warburg, 1980); H. Abbott, 'Do the Little Things Right Every Day' (1982) 4 *Prod. Liab. Int.* 82–3.

[6] L. Preger *et al.*, *Asbestos-Related Disease* (NY, Grune & Stratton, 1978), 121–2.

[7] See, respectively, D. Kent and L. Cenci, 'Smoking in the Workplace: Tobacco Smoke Health Hazards to the Involuntary Smoker' (1982) 24 *J. Occup. Med.* 469; J. Robinson, 'Cancer of the Cervix' in A. Jordan, F. Sharp, and A. Singeas (eds.), *Preclinical Neoplasia of the Cervix* (London, Royal College of Obstetricians and Gynaecologists, 1982), 11.

usually resulting in immediately palpable injury. Medical cause in fact is rarely a problem. In contrast, non-traumatic disability cannot be attributed to a particular and isolated point in time and place. This is either because it is the result of a 'cumulative trauma' such as the pneumoconioses (due to dust accumulation in the lungs) or because the time of injury cannot be determined as in cases of infection. This problem of real or apparently insidious contraction renders identification of the potential defendant extremely difficult in non-traumatic disease cases. When combined with the characteristic and variable delay in the appearance of symptoms (latency) in such cases, the result is that it is nearly always impossible to establish that a certain defendant's action was medically a cause in fact of the injury. Moreover, even if this were possible, the uncertainty of when harm was done and the passage of time before symptoms appear mean that evidence of fault is often unavailable.

The second complicating characteristic of non-traumatic disease is the problem of multiple causation. Unlike the case in accidents, the causal factors contributing to the disability may be unclear. For example, the disease may be one which sometimes occurs from man-made causes and sometimes from other or uncertain causes. An example of this 'multiple *possible* causation' problem is lung cancer. How can it be shown that the man-made cause operated in an individual case? Another multiple causation problem is that the individual case may be attributable to more than one cause. Examples of this 'multiple *cumulative* causation' are the pneumoconioses where successive periods of exposure to airborne dust contributed to the resultant disease. Two unique factors complicate disease cases at this stage. First, one of the multiple factors which contributed to the disability may be the physiological susceptibility of the victim to the hazard. Clearly this is rarely a problem in accident cases: if hit by a large rock most miners would suffer traumatic injury, but in the mining work-force only some will become disabled by the dust all inhale. Secondly, in cases of multiple cumulative causation, two factors may be multiplicative (that is, 'synergistic') in their effect. These characteristics mean that the discovery of medical causation and then its proof in individual cases is extremely difficult, while they also introduce complex questions about apportionability both over time and between human agencies.

1.4 *Compensation for Man-Made Disease*

In England victims of man-made diseases are usually in no better position than victims of any other disease. In both cases the victims are unlikely to be covered by private health insurance. In both cases the victim, if an earner, may have limited rights to statutory sick-pay in the short term, and, in some cases, some form of voluntary arrangement with the employer may provide additional relief. However, the relief available to disease victims—especially those with long-term disability—is usually limited in practice to that provided by the social security system regardless of whether the disease was man-made.

None the less, in theory, the victims of man-made disease may have two additional remedies, and it is with these and their failure in practice with which the initial chapters of this book are concerned. Firstly, the victim may qualify for benefit under the industrial injuries scheme. This remedy, now limited to cases of death and longer term disability, is available where the victim suffers from one of a list of prescribed industrial diseases and can satisfy certain qualification requirements. To give an idea of the incidence of these benefits: in the year 1981–2 prescribed industrial disease other than pneumoconiosis and byssinosis formed the basis of 7,210 awards of which 6,379 were spells of industrial injury benefits (that is, short-term injury); there were 891 new cases of pneumoconiosis and byssinosis diagnosed by pneumoconiosis medical panels; 884 awards of industrial death benefit were made on the basis of prescribed disease; and persons receiving benefits for pneumoconiosis and byssinosis numbered approximately 29,700. (In contrast, accidents accounted for 556 awards of death benefit and 384,000 spells of industrial injury benefit.)[8] The principal advantage the scheme has over tort (and industrial compensation systems such as those in the United States and Australia, which are based on the liability of a specific employer) is that fault in an identifiable wrongdoer need not be shown, nor, in most cases, need the claimant affirmatively prove medical causation, as he or she can take advantage of presumptions to this effect. The principal advantages over social security benefits are that industrial benefits are more generous and they are incapacity-based, so that

[8] Health and Safety Executive, *Health and Safety Statistics 1981–1982* (London, HMSO, 1985).

total incapacity for work is not a prerequisite as it is under the general scheme. The major disadvantage is that the scheme is restricted to certain diseases omitting what has been roughly estimated as 80 per cent of occupational disease cases.[9] The net result is that this remedy is formally biased towards traumatic injury.

In contrast, the second additional remedy which may be available to the victim of man-made disease—tort—is formally available whether or not the disease is traumatically caused and whether or not the disease was caused by the victim's work. However, in addition to the problem of medical causation which still remains, there is the disadvantage of having to identify a particular individual and (usually) to prove the injury was due to his fault. From the surveys conducted by the Pearson Commission it is known that only 6.5 per cent of non-fatal 'injuries' attract some damages in tort.[10] Even though they form a minority of those injured most tort claims concern the victims of road accidents (41 per cent) and work-related causes (47 per cent), both of which are disproportionately successful.[11] No figures exist on the number of tort claims made in relation to non-traumatic injury, but from reported cases and the literature it appears clear that very few such claims are made and for these the success rate is low. This is confirmed by the relative claim rate and success of occupational accident and occupational disease claims: reflecting the general pattern just noted in which occupational claims are disproportionately represented, occupational *disease* claims form the majority of non-traumatic claims made in tort. This is accounted for not only by the presence of compulsory employers' insurance and the power of organized backing through unions but because of the much greater possibility of identifying the potential defendant and establishing medical causation (see section 3.2). None the less, even in the occupational context disease claims are relatively rare—of the 47 per cent of tort claims based on work-related causes, only 1.2 per cent concerned occupational disease, and the final success rate is lower, with only 57 per cent of occupational disease claims in tort resulting in some payment, while 78 per cent of occupational accident claims result in payment.[12]

[9] Pearson, vol. 2, para. 173.
[10] Pearson, para. 78.
[11] Pearson, paras. 37, 78–9.
[12] Pearson, vol. 2, Table 19; Pearson, vol. 2, para. 180.

Turning to the reported case-law, it is now well recognized that since only 1 per cent of tort claims reach trial and even fewer are reported, the pattern of reported case-law is unrepresentative of claims being made in practice. Thus, for example, deafness, one of the most common occupational diseases, has given rise to thousands of successful common law claims but only a few have been reported in which liability was an issue. Even more striking is the paucity of reported English case-law on the issue of liability for asbestos-related disease or byssinosis despite the fact that these diseases have generated thousands of civil claims. Clearly then, the dearth of reported disease case-law overall is not just attributable to the relatively low claim rate compared with accident claims. A particular factor preventing disease cases getting to court—one which goes to explain the low rate of reported cases even in high claim areas—may be the much greater 'ripple' effect a finding of medical causation or negligence has in these contexts. Except in the occasional mass disaster situation such as an aircraft crash, this is not an important factor in accident case-law. In the disease context, however, it is typical for many claimants to have broadly equivalent claims. They have, for example, taken the same pharmaceutical or worked under the same hazardous industrial conditions. Defendants may consequently be keener to avoid trial and the possibility of a damaging and public precedent with its flow-on effect (see section 5.6). (It is unclear whether this would produce a relatively greater pressure to settle out of court or whether settlements operate as precedents in much the same way as court judgments, with the result that defendants are relatively more resistant to disease claims even in the settlement context.) Conversely, where there is some sort of organization backing plaintiffs such as a union, the attraction of a test case or test settlement establishing such a precedent is great, and has been used to press for the creation of new compensation schemes such as that covering coal miners' pneumoconiosis, and the expansion of existing schemes such as the scheduling of occupational deafness under the industrial injuries scheme.

In the more litigious context of the United States, man-made disease cases—variously described as 'cumulative trauma' or 'toxic torts'—are much more numerous. Asbestos litigation alone—more than 30,000 claims were pending against 260 firms in 1983—it has been said, 'threatens to swamp the courts and to bankrupt defendants' and their insurers.[13] Judges have been appointed to work full time on complex asbestos cases, and the largest asbestos firm in the western world has filed for bankruptcy in recognition of its forecast liability in the US for asbestos diseases.[14] But even in the United States causation problems have led to what one commentator described as 'an almost complete breakdown in the tort system as a mechanism for internalizing . . . costs'.[15] Indeed, only half the asbestos claims which get to US courts succeed, and, according to the Chairman of the Lloyds Underwriters' Association, Mr Richard Ballantyne, more is being spent on lawyers' fees than on damages to successful asbestos claimants.[16]

1.5 *Outline*

The overall prospects for a victim of man-made disease obtaining tort compensation are grim. The conceptual requirements of the remedy ensure that in practice in England it has been almost exclusively restricted to the victims of accident, and this is reflected in the preoccupation with traumatic injuries in academic and reform literature to the neglect of issues raised by disease cases. Chapters 2, 3, and 4 look at the three principal areas of conceptual difficulty: problems relating respectively to time, causation, and the proof of fault. Comparable difficulties in the other major compensation system, the industrial injuries scheme, are analysed when appropriate. Chapter 5 looks at possible reforms of the common law remedy via stricter forms of liability and the viability of reform by limited non-tort systems of compensation. Chapters 6 and 7 look at two policy issues: non-compensatory goals, specifically economic deterrence (Chapter 6); and the implications for the reform agenda by the consideration of the issues raised by disease (Chapter 7).

In short, the book re-evaluates, from the neglected perspective of the disease victim, criticism of the tort remedy and the limited

[13] L. Treiger, 'Relief for Asbestos Victims: A Legislative Analysis' (1983) 20 *Harv. J. on Leg.* 179–82; R. Vagley and M. Blanton, 'Aggregation of Claims' (1981) 16 *Forum*, 636, 648.

[14] R. Epstein, 'Manville: The Bankruptcy of Product' (Sept.–Oct. 1982), *Regulation* 14.

[15] R. Pierce, 'Encouraging Safety: The Limits of Tort Law and Government Regulation' (1980) 33 *Vand. LR.* 1281, 1298.

[16] E. Anderson, I. Warshauer, and A. Coffin, 'The Asbestos Health Hazards Act' (1983) 10 *J. of Legislation*, 25, 32; 'World Insurers Pay out £58,000 million for Asbestosis', *The Times*, 27 Aug. 1982, 13.

reform debate to which it has given rise. Broadly it could be said to start from Calabresi's question: 'Why is compensation for illness even in highly welfaristic countries much less complete than compensation for accident victims?'[17]

[17] G. Calabresi, *The Costs of Accidents* (New Haven, Yale Univ. Press, 1970), 43.

2

Time and Disease

2.1 *Time Characteristics of Disease*

Many of the legal problems associated with claims for compensation for man-made disease flow from two time characteristics of such diseases: gradual contraction and latency. Although diseases such as cancer may be caused by trauma it is more typical for man-made diseases to be gradually contracted over time and it is with these diseases that this book is concerned. In some cases, such as fibrosis of the lung, a long period or high level of exposure to the hazard is necessary to produce disease. In other cases, severe disease can result from quite short and low-dose exposure, a notable example here being mesothelioma, an asbestos-caused cancer which apparently is invariably fatal. What these diseases have in common is that they cannot be ascribed to a single isolated trauma so that their contraction and insidious progress may go unnoticed by the victim until later diagnosis. This distinctive 'gradual-contraction' characteristic poses the most fundamental dilemma for traditional legal concepts, because, even when the disease is diagnosed, it is not possible to pin-point a single time and place when damage was done.

No less important but more practical difficulties are created by the second time characteristic associated with many man-made diseases: the delayed manifestation of symptoms or 'latency'. In some cases symptoms may appear soon after exposure commences but in others there is a considerable latency period between the initial damaging exposure and the appearance of symptoms. Thus, man-made diseases now being seen may be the result of exposures in the 1930s, 1940s, and 1950s, while present day exposures will be reflected in disease not manifesting itself until next century. The phenomenon can be illustrated by the case of a secretary who worked for a company manufacturing asbestos products in the United States from 1940 to 1955.[1] Twenty years after leaving this

[1] *Harig* v. *Johns-Manville Prods. Corp.* (1978) 394 A. 2d 299.

employment she developed a cough and a year later, in 1976, was diagnosed as suffering from mesothelioma. As earlier noted, this disease may be contracted by very short or low-dosage exposure, which explains why family members of asbestos employees and workers not directly using asbestos, like the secretary, are known to contract this cancer. The fact that latency periods of up to fifty years have been recorded for man-made diseases gives some indication of the problems facing victims of diseases such as mesothelioma when they seek compensation. These include procedural and entitlement requirements based on time such as limitation periods and—in industrial compensaton systems—minimum or recent exposure rules, as well as conceptual difficulties such as the meaning of 'damage' and its apportionability, and it is these that this chapter addresses. There are other problems stemming from the characteristics of gradual contraction and latency, such as problems in proving medical causation and negligence long after the relevant exposure, but these are dealt with in later chapters.

There are three types of legal requirements relating to time which may face a disease victim seeking compensation. Firstly, there may be a conventional limitation period within which a claim must be brought. This is the only time condition affecting common law claims. But within the industrial compensation schemes (the workmen's compensation scheme and its replacement, the industrial injuries scheme) there also are, or have been, 'recent exposure' and 'minimum exposure' rules which predicate entitlement to compensation on the victim having been exposed to the relevant hazard within a certain period before making the claim or for a certain minimum length of time. In general, these rules were developed in an attempt to increase the likelihood that the cause of the successful claimant's disease lies within the causal limitations of the scheme (for example, that it was related to a work process) and was not due to some extraneous source. Each of the three types of legal requirement relating to time have inhibited disease claims in one way or another and this has often been neglected when reform of the law has been considered. Since historically industrial compensation schemes from 1906 provided the first significant avenue of recovery of compensation for disease victims—in this case victims of certain listed occupational diseases—it is useful to start with the time requirements under these systems and the problems they revealed before a description of the situation at common law is made. What will be seen is that many of the time problems facing disease victims seeking compensation in tort had become evident from an early date under industrial compensation schemes. Set in this context, the fact that the common law rules relating to time were not reformed until the 1960s is shown to be quite extraordinary and is a powerful illustration of the neglect of disease issues.

2.2 *Time Barriers Under Industrial Compensation Legislation*

(*a*) *Limitation Periods* Under the industrial compensation schemes the potential problems relating to time involved in the recovery of compensation for disease had become manifest at an early stage. Firstly, under the workmen's compensation (and later the industrial injuries) scheme there has always been a provision which allowed claims brought outside the normal limitation period to proceed if the delay in claiming had been due to a 'reasonable' or 'good' cause.[2] What this means in practice is that a claimant who, because of a reasonable excuse, did not claim within the limitation period running from the date of first entitlement to claim is not barred. Under the current system this is a particularly useful rule in claims for disablement benefit because, since impairment of earning capacity is not a prerequisite of entitlement for this benefit, it is potentially payable from a much earlier date in the course of the disease (now set at 90 days after the 'date of onset', that is, of loss of any physical or mental faculty) than the now abolished injury benefit which was predicated on total disablement from work. Where the delay in claiming disablement benefit is held to be reasonable, full backdating to the often remote date of onset is allowed.

There are particular reasons why a victim of disease may not claim within a limitation period, reasons which would rarely if ever apply to accident claims. Firstly, there is the obvious problem of misdiagnosis (e.g. R(I) 6/54). Since a disease victim will not have suffered a manifest trauma with which to associate his disablement, he may not connect it with his work environment, and his doctor may be similarly unable to identify the victim's condition as work-related. By the time the work-relatedness of the disease is

[2] Workmen's Compensation Act, 1906 (6 Edw. 7, c. 58, s. 2(1)); Social Security (Claims and Payments) Regulations 1979, SI 1979, No. 628, Reg. 14.

diagnosed the limitation period may have elapsed. Secondly, even when the doctor realizes the disease is work-related, he may give the patient a euphemistic diagnosis to minimize anxiety—as when a doctor tells an asbestos worker his disease is asbestosis (a long-term degenerative disease) when in fact it is mesothelioma (a rapid and fatal cancer). This common and government-sanctioned practice may lull the patient into a belief that there is less urgency in claiming promptly than is the case (see Chapter 3). Thirdly, the medical adviser may tell the patient that he has a work-related disease which is one of those covered by the industrial compensation scheme but wrongly advise the patient that it has not yet produced a compensatable level of disability (for example, R(I) 40/59). These sorts of characteristic reasons for delay in claiming for disease—rarely applicable, if ever, to accident cases—were well known and well accommodated within the industrial compensation system because it contained from the start a provision allowing claims made outside the limitation period where the delay was for a 'reasonable' cause.

There was and remains, however, one problem area. It is a general rule that ignorance of the law is no excuse. This means here that if a worker knows that his injury is work-related, it is no excuse to say that he did not claim within the limitation period because he did not realize that such an injury gave him a legal entitlement to compensation (see e.g. R(I) 82/53). Whatever one may think of such a general rule, it often has a particularly harsh application to disease victims. This is because the industrial compensation system in Britain is, and always has been, a typical non-comprehensive compensation system in that it focuses on accidents and is highly selective as to which disease claims are to be covered. The result is a much lower consciousness of the possibility of a disease claim as opposed to one for traumatic injury. As Commissioner Lazarus noted in R(I) 14/74,

> It is expecting a lot of people that they should suspect that a scheme mainly designed to deal with accidental injury and described as an injury scheme may also deal with diseases which develop otherwise than by accident and are not injuries in the sense in which that word is commonly used.

(*b*) *Minimum Exposure Rules* As Chapter 3 will describe, one of the greatest problems facing a disease victim is proof of medical causation. One way designers of industrial compensation systems have in a few instances sought to increase the systems' coverage of diseases is to predicate a rebuttable presumption of causation (that is, of work-relatedness) or entitlement itself[3] on what are called 'minimum exposure rules', a requirement that the claimant had been exposed to the prescribed conditions for a minimum period of time. Since the design of the scheme is that coverage should only extend to disease cases where it could be assured to a fairly high degree of certainty that the disease was work-related, the idea is that by requiring certain minimum exposure periods this could be achieved for a wider range of cases. But despite this positive role in facilitating compensation for certain diseases, from the start of the industrial compensation scheme just after the turn of the century, these requirements excluded, and were seen to exclude, many otherwise good claims because the victim had less than the minimum exposure. Moreover, minimum exposure requirements —especially when mandatory, as they were in the case of the cotton dust disease byssinosis and still are for occupational deafness—hinder prevention by indirectly encouraging workers to continue at work until the period has been served.

(*c*) *Latency and Recent Exposure Rules* One of the most striking problems facing a disease victim is the latency of the disease, that is where symptoms do not appear until a considerable time after the relevant exposure. From the start of the industrial compensation system it was clear that latency was a potentially major problem with disease claims. In the case of the requirement to claim within the limitation period, the problem was not significant because benefits were, and still are, predicated on disability and, inasmuch as latency of symptoms exacerbated the problem of recognition of work-relatedness, it was accommodated within the 'reasonable cause' for delay provision. However, in one important context latency was a dramatically visible barrier to disease claims within the industrial compensation system, and that was in cases where 'recent exposure rules' applied. A workmen's compensation claim

[3] See Social Security (Industrial Injuries) (Prescribed Diseases) Regulations 1980, SI 1980, No. 377, Reg. 4(3) (non-mandatory two-year minimum exposure rule for rebuttable presumption of work-relatedness in cases of pneumoconiosis); Social Security (Industrial Injuries) (Prescribed Diseases) Amendment (No. 2) Regulations 1983, SI 1983, No. 1094, Reg. 3(1) (mandatory ten-year minimum exposure rule for deafness).

could only be made where the victim's scheduled disease was due to his employment within twelve months immediately prior to the date on which the worker became sufficiently disabled to be entitled to compensation under the statute (that is, 'statutory disablement').[4] The purpose of this requirement of recent exposure was to accommodate diseases of gradual contraction within a scheme designed to impose liability to pay compensation, at least initially, on a single identifiable employer. But this accommodation within a scheme fundamentally intended and designed for the compensation of accidents was inevitably crude and awkward. In operation it excluded *ab initio* all claims where the date of last exposure to the hazard was more than twelve months before the date on which the worker became statutorily disabled. The worker may have moved to a different sort of job before any symptoms appeared. Even more disturbing was the possibility that if a worker began to exhibit signs of illness he may have moved or been moved by his employer to a safer job. He may simply have been sacked. If the progress of the disease did not then result in statutory disablement until more than a year later, a claim could not be made. The rule not only encouraged employers to dismiss sickly workers in an attempt to avoid paying compensation; it also indirectly encouraged sick workers to stay in dangerous working conditions until they were statutorily disabled.

Although the problem was noted in 1909 by an early commentator on the scheme and quickly illustrated by the failure of early industrial disease claims,[5] it did not prevent the adoption of recent exposure rules (designed in this case to increase diagnostic certainty) in the special compensation schemes set up after 1918 to cover certain progressive lung diseases.[6] Despite some criticism in Parliament these mandatory recent exposure requirements survived, and it was not until the replacement Industrial Injuries scheme was being formulated in the early to mid-1940s that criticism of the rules generated sufficient pressure for their removal.[7]

[4] Workmen's Compensation Act 1906, s.8(1).

[5] E. Lawes, *The Law of Compensation for Industrial Diseases* (London, Stevens & Sons, 1909), 29–31; *Hughes* v. *Harrison* (1931) 24 BWCC 104.

[6] e.g. Metal-Grinding Industries (Silicosis) Scheme 1927, SRO 1927, No. 380.

[7] A mandatory recent exposure rule (5 years) still applies in the case of deafness, SI 1983, No. 1094, Reg. 8, in order to limit claims, and a non-mandatory recent exposure rule (1 month) still applies to the rebuttable presumption of work-relatedness in the general case: SI 1980, No. 377, Reg. 4(1).

It could be argued that, given the low priority accorded the problems of disease claims as opposed to accident claims, abolition of this barrier may have taken much longer had it not been facilitated by the dramatic and far-ranging social reforms involved with the Beveridge proposals—it being easier to incorporate reforms in a scheme *ab initio*. Certainly in jurisdictions such as the United States which retain the traditional workers' compensation system, mandatory recent exposure rules survive criticism to remain a significant barrier to disease claims.[8] On the other hand, even in New Zealand recent exposure requirements survived the unique and fundamental reform of personal injury compensation in the early 1970s and have only recently been abolished,[9] perhaps indicating the extent of the neglect of disease issues by formulators of that scheme.

2.3 *Common Law Claims, Statutes of Limitation, and Latent Damage*

This short sketch of the operation of the time rules under the industrial compensation schemes reveals that from an early date—even as early as before the First World War—certain characteristic limitation problems facing the disease victim, such as latency and misdiagnosis, would have been apparent to the interested observer. This makes all the more remarkable the fact that until 1963 most common law disease claims remained wholly or partially barred by an *absolute* statutory period of limitation. At that date this stood at three years from the date the cause of action accrued which, in claims of negligence or breach of statutory duty, is when damage occurs. For common law claims there was no relief granted by a 'reasonable cause for delay' provision so that the fact that a claim was not brought within three years of accrual because of misdiagnosis or latency of symptoms would not save it.

Despite the lengthy experience of such problems from the industrial compensation context, the failure of a number of reported common law disease claims on the ground of being out of time, and despite the considerable exposure of the issues in academic debate and reported case-law in the United States

[8] P. Barth and H. Hunt, *Workers' Compensation and Work-Related Illnesses and Diseases* (Cambridge, MIT Press, 1980), 122 Table 4.3.

[9] Accident Compensation Act 1972 (NZ) (1972, No. 43, ss. 67(1) and 68(1)); Accident Compensation Act 1982 (NZ) (1982, No. 181, ss. 28(1) and 29(1)).

(including a leading Supreme Court decision on the point in 1948, *Urie's* case), the limitation problems of the disease victim were not widely recognized in Britain.[10] They did not even attract the attention of law reformers concerned with limitation periods, even when specifically directed to the industrial compensation context where the issues were well known.[11] Reform was not precipitated here until the criticism following the Court of Appeal decision in *Cartledge* in 1961 (later upheld on appeal to the House of Lords in 1963).[12] This decision held that the plaintiffs who were suffering from pneumoconiosis from their previous working conditions were statute-barred from suing the defendants because their cause of action had accrued more than the relevant limitation period before the date of claim. The fact that the plaintiffs' symptoms had not become reasonably discoverable until the limitation period had already expired was regarded as irrelevant. It was held that one could suffer damage and therefore have a cause of action in tort yet be unaware of it and the limitation period could therefore elapse before one is aware of any damage and in a practical position to sue.

The reasoning in the judgments in *Cartledge* was unsatisfactory in a number of ways (see 2.4 below), but what it achieved was to focus attention at long last on the injustice of a fixed limitation period in the context of disease damage, in particular on the problem of latent damage. The result was a fundamental change in the foundation of statutory limitation law as it applied to claims for personal injuries. The Limitation Act 1963 began a series of awkward, although ultimately successful, statutory reforms to ameliorate the situation. Henceforth, time would not begin to run against the victim until he was, or should reasonably have been, aware of certain material facts. That the victim was suffering damage was one of the most important facts, so that time would not run whilst the disease was latent. Another vital material fact

[10] See e.g. *Archer* v. *Catton & Co. Ltd.* [1954] 1 All ER 896; *Clarkson* v. *Modern Foundries Ltd.* [1958] 1 All ER 33; *Coote* v. *Eastern Gas Board* [1953] 1 QB 594. On the US academic debate see Anon., 'Developments in the Law; Statutes of Limitation' (1950) 63 *Harv. Law Rev.* 1177, 1203–7, and the review of case-law by W. Allen, 'Limitations: Contracting Disease' (1950) 11 ALR 2d 277. See also *Urie* v. *Thompson* (1948) 337 US 163.

[11] See *Final Report of the Departmental Committee on Alternative Remedies* (1946) Cmd. 6860.

[12] *Cartledge* v. *Jopling & Sons. Ltd.* [1961] 2 Lloyds Reps. 61; [1963] AC 758.

was the medical cause of the disease: now the mere fact that the victim knew he was diseased would not start time running against him because a material fact was the accurate medical diagnosis of the disease as possibly due to the relevant man-made cause, for example work conditions in an occupational disease claim or a particular pharmaceutical in the case of a claim for an adverse drug reaction.

One problem persisted. As described earlier, because of the paucity of reported disease claims and litigation it is more likely that a disease victim will be ignorant of his potential legal entitlement even when he knows that his medical condition was due to a particular medical source. However, under the Limitation Act 1963 it was unclear whether knowledge that the condition and its cause gave rise to a legal cause of action was a material fact. Eventually the question was resolved in the negative, creating problems for disease victims who, even if they knew their disease was attributable to a man-made cause, did not suspect they might have a cause of action.[13] Even this difficulty has since 1975 been overcome by the 'revolutionary' (to use Lord Denning's word) inclusion within the limitation legislation of an overriding judicial discretion which allows the court—if it is equitable to do so—to permit claims brought out of time even if the delay was not due to ignorance of a material fact.[14] Now, for example, delay due to ignorance of legal entitlement may not be fatal to the claim. Similarly, victims of progressive diseases, who know of their potential cause of action, but who for an initial period prefer not to pursue their legal claim and who only later (often more than a decade later) as the disease progressively worsens launch a claim, have had the discretion exercised in their favour, the courts openly approving their initial conciliatory non-litigious inclinations. In one particular case a delay of fifteen years was excused despite 'real prejudice' to the defendant in having to recall conditions fifteen to twenty years ago.[15] Clearly then, the problems of the victim of disease damage have prompted a fundamental shift in limitation law, overriding the traditional justification for absolute

[13] *Central Asbestos Co. Ltd.* v. *Dodd* [1973] AC 518.

[14] *Firman* v. *Ellis* [1978] 2 All ER 851, 859; Limitation Act 1980 (c. 58, s. 33).

[15] *Brooks* v. *J. P. Coates Ltd.* [1984] 1 All ER 702. See also *Buck* v. *English Electric Co. Ltd.* [1977] 1 WLR 806; *McCafferty* v. *Metropolitan Police District Receiver* [1977] 2 All ER 756.

fixed limitation periods (i.e. prejudice to the defendant in having to face claims, even if justified claims, many years after the relevant events). The regime now accommodates the typical reasons for disease victims not claiming promptly: latency of symptoms; the degenerative ill health itself; euphemistic or incorrect medical advice; inadequate or inordinately discouraging legal advice from lawyers inexperienced in such an underlitigated area as disease; difficulty in identifying the potential defendant; a belief that compensation being received from one source—typically from the Industrial Injuries Fund—was all the victim was legally or morally entitled to;[16] and a perception that, whereas work accidents may be the legal responsibility of the employer, the risk of disease is one inherent to the conditions of employment and perceived as one for which the employer is not responsible.[17] Finally, it must be stressed that these factors which the limitation cases reveal result not only in delay in claiming but far more often are the reason why disease victims never even seek advice as to their common law rights and never enter the tort litigation system at all. The question is returned to in Chapter 5.

2.4 *What is Latent Damage?*

A pivotal conceptual question underlies limitation law. This is: what is the damage which grounds the action? This is, or at least was, important in limitation questions because the cause of action does not accrue until the damage was suffered. Formally time then ran from the date of damage. So it is important to know what courts regard as the damage in cases like latent disease. If it is concluded that the intrusion into the body of foreign matter such as chemicals, dust, or X-rays, without any additional cause save the victim's own physiological responses, triggers deleterious bodily changes that eventually produce manifest disability and that this process must, however microscopically, have commenced at the time of intrusion, then damage will be constituted by these microscopic changes, and occur contemporaneously with the intrusion of foreign matter, that is, with exposure. This means that the tort is a continuing tort and will give rise to a fresh cause of

[16] e.g. *Central Asbestos Co.* v. *Dodd* [1973] AC 518; *Skingsley* v. *Cape Asbestos Co. Ltd.* [1968] 2 Lloyds Reps. 201; *Harper* v. *National Coal Board* [1974] 2 All ER 441; *Newton* v. *Cammell Laird & Co.* [1969] 1 WLR 415.

[17] *Newton* v. *Cammell Laird & Co.* [1969] 1 WLR 415, 421.

action *de die diem* while exposure lasts; that the latest limitation period runs from the date of last exposure; and under a fixed limitation period of, say, three years the plaintiff could only sue for the damage due to exposure on days within three years of the writ, not before.[18] If there was no exposure during this time, that is, the date of last exposure was more than three years before the date of writ, the claim would be totally barred. [Clearly then, one of the practical advantages of the limitation reforms, especially the judicial discretion, is that it not only allows what might otherwise be completely barred claims, but also allows *full* recovery for any claim that it allows through including claims which would otherwise be only partially barred because some of the relevant exposure fell within the formal limitation period.[19]]

The major alternative theory is that damage does not occur, and therefore the cause of action does not accrue nor the limitation period run until the injury was discovered or reasonably discoverable. This 'discovery' theory appealed to many as an attractive device to assist victims of latent damage to overcome fixed limitation periods. Despite the conceptual difficulty of seeing damage as dependent on knowledge, this discovery theory was adopted by the Supreme Court in the United States as early as 1948 in an occupational disease claim and it remains a dominant, although by no means universal, interpretation of limitation questions in that country.[20] It has even enjoyed a brief ascendancy in this country in cases of latent physical damage. None the less, in cases of personal injuries the discovery rule has never been adopted by a superior court in England and was explicitly rejected by the House of Lords in *Cartledge* (and now in *Pirelli* for cases of physical damage[21]). According to the Law Lords in *Cartledge*, damage does not depend on knowledge or discoverability. Indeed, although not necessary for the decision, the Lords clearly thought damage could occur *before* it was discoverable by *any* means at an earlier indeterminate date they called the time of 'the secret onset of the disease'. The difficulty is that the House of Lords gave no clear statement of what they recognized to be the damage in cases

[18] *Smith* v. *Central Asbestos Co. Ltd.* [1971] 3 All ER 204, 208; *Cartwright* v. *G. K. N. Sankey Ltd.* [1972] 2 Lloyds Reps. 242, 248.

[19] See e.g. *Smith* v. *Central Asbestos Co. Ltd.* [1971] 3 All ER 204.

[20] *Urie* v. *Thompson* (1948) 337 US 163; Anon., 'Preserving Causes of Action in Latent Disease Cases' (1982) 68 *Virg. L. R.* 615.

[21] *Pirelli General Cable Works Ltd.* v. *Oscar Faber & Partners* [1983] 2 AC 1.

of latent damage. Specifically, they did not adopt the simple and arguably common-sense notion that damage occurs contemporaneously with exposure, for if they had there would have been no need for a concept of 'secret onset' and the latest limitation period would simply have run from the date of last exposure (see above). This major gap in reasoning means we not only do not know when damage occurs (how to pin-point when 'secret onset' occurs) for the purposes of limitation law—admittedly, no longer a barrier to claims because of the statutory adoption of the discovery rule from 1963 onwards—but what the damage is and when it occurs for the purpose of apportionment between tortfeasors and for the purpose of insurance cover (the latter being a source of intense and very costly litigation in the United States).[22] Needless to say, equivalent problems do not arise in accident cases because there the trauma almost invariably results immediately in some recognizable significant injury such as a scratch or broken leg, which is clearly damage and gives rise to a cause of action at that time, even though much worse bodily injury might become apparent later.

In practice the question of what the damage is in latent disease cases is glossed over by the *de facto* adoption by courts and insurers of the exposure rule—that damage occurs contemporaneously with exposure and so can be divided, admittedly in only a crude way, on the basis of responsibility over time.[23] This means for example that, where exposure to a particular hazard was regarded as negligent only after a certain date, the plaintiff recovers damage only on the basis of aggravation after that date and before the date, if any, when such negligent conditions ended.[24] The same approach is applicable where there are

[22] Anon., 'Adjudicating Asbestos Insurance Liability' (1984) 97 *Harv. L. R.* 739; J. Ingram, 'Insurance Coverage Problems in Latent Disease and Injury Cases' (1982) 12 *Environ. Law*, 317.

[23] e.g. *Ransom* v. *Sir Robert McAlpine & Sons Ltd.* (1972) 11 KIR 141, 163; *Berry* v. *Stone Manganese and Marine Ltd.* [1972] 1 Lloyds L. Reps. 182; *Smith* v. *Central Asbestos Co. Ltd.* [1971] 3 All ER 204, 208; *Cartwright* v. *G. K. N. Sankey* Ltd. [1972] 2 Lloyds Reps. 242, reversed on the question of liability: (1973) 14 KIR 349; *Thompson* v. *Smiths Shiprepairers Ltd.* [1984] 2 WLR 522. On insurers see E. Whitmore, *Employers' Liability Insurance* (London, Pitman, 1962), 82; P. Neild, *Byssinosis: The Lancashire Disease* (London, Chartered Insur. Instit., 1982), 42.

[24] *Crookall* v. *Vickers-Armstrong Ltd.* [1955] 2 All ER 12; *Wallhead* v. *Ruston & Hornsby Ltd.* (1973) 14 KIR 285; *Thompson* v. *Smith Shiprepairers Ltd.* [1984] 2 WLR 522.

sequential wrongdoers (and/or insurers), a typical problem in cases of diseases of gradual contraction. But we still do not know, for example, what approach will be taken and why when a single tortfeasor has control over a number of sources of risk, say of machines producing dust in a factory, which contemporaneously cause damage but where only one source of dust was operated negligently. Should the victim recover his full damage or should it be only for the proportion of damage calculated as due to the negligent source?[25] If damage is regarded as the microscopic effects occurring simultaneously with exposure and therefore divisible over time it seems logical that it should be treated as divisible with respect to sources (and this also has implications for the question of whether in a disease case contemporaneous tortfeasors are independent or concurrent). The problem is in all these issues of apportionment that there has been no clear statement and justification of the principles to be applied. This situation and the fact that the reasoning in *Cartledge* was so vague and inadequate on the concept of damage strikingly show how poorly courts have handled many of the novel conceptual problems presented by disease cases. We see this yet again in issues of causation (Chapter 3).

2.5 *Conclusion*

To sum up: of the three types of time rules which might confront the victim of man-made disease seeking compensation, mandatory recent and minimum exposure rules were in the past incorporated only into systems of compensation for occupational disease. The primary reason for such rules was to improve diagnostic certainty, but they resulted in many disease claims being barred. The third type of time rule is the limitation period—the requirement that a claim be made within a certain time of injury. The traditional rationale of such a rule is threefold:[26] it encourages diligence in bringing actions promptly; it recognizes the fact that evidence usually becomes stale and unreliable with the passage of time; and it attempts to provide defendants with a degree of personal certainty that they will not be left open to liability indefinitely. For

[25] Compare *Cartledge* v. *Jopling & Sons Ltd.* [1961] 2 Lloyds Reps. 61, 69–70 (per Harman LJ); 79 (Pearson LJ).

[26] *Report of the Committee on Limitation of Actions in Cases of Personal Injury* (1962) Cmnd. 1829, para. 17.

a number of reasons these rules operate particularly harshly in disease cases where delay in making a claim is often the entirely reasonable result of unknown medical causation, misdiagnosis or, most importantly, of latency of symptoms themselves. In the state compensation system a principle of 'reasonable cause for delay' has ameliorated most of these problems. With common law claims, however, strict principles of statutory interpretation and inadequate concepts of 'damage' prevented courts fashioning an equitable solution. Statutory reform of this area of the law was remarkably late in England—a striking example of how little may be learnt from the experience under the state industrial compensation systems and from other jurisdictions.

When the attention of reformers was at last attracted to the problems of disease victims, a dramatic and far-ranging shift in policy was provoked. Henceforth it was recognized that delay may be excused because the prejudice to the plaintiff outweighed the prejudice to the defendant. In consequence, statute recognized a discovery rule which ran counter to both the personal certainty and evidentiary arguments protecting the defendants' interest. Those supporting reform noted that there was little difference in the threat posed to the defendants' sense of certainty by such a rule and that which already existed in cases where the actual personal injury occurred long after the relevant wrongful act, such as where a latent defect in a tool caused traumatic injury years after negligent manufacture. Moreover, although it is true that proof becomes more difficult with the passage of time, this usually presents much greater difficulties to the plaintiff who has the onus of proof, particularly in disease cases (see Chapters 3 and 4). Despite complexities of drafting, the statutory reform has proved to be a better solution than the attempts at reform by courts in the United States.

On the other hand, the process of reform also shows how persistent the neglect of disease issues has been and continues to be. For, not only was recognition of the limitation problem of latent disease extraordinarily late in being taken, but it could well be argued that it has been both superficial and its lessons poorly thought out. One example of the latter is the failure to integrate in any adequate sense the development marked by the *Cartledge* decision into limitation rules for latent damage outside the personal injuries field. As noted earlier, in the past decade even the discovery rule itself—so clearly rejected by the Law Lords in 1963 on basic principles—enjoyed a brief ascendance in the area of actions for physical loss—indeed, more than twenty years passed before the implications of *Cartledge* were recognized here and adjustment made in doctrine to bring it into line with that decision.[27] Secondly, and even more remarkable than this failure to apply the lessons of latent disease in the wider context, is the minor importance they are now accorded even in the context of limitation rules for personal injury. The example here is product liability reform where proposals to introduce strict liability for defective products contain a requirement that any such claim be brought within ten years from the date the product was put into circulation.[28] One of the purposes of the scheme is to overcome the difficulties of proof of negligence facing those injured by defective goods produced under complex design and manufacturing processes. That being the case, the problems of this sort encountered by the victims of latent damage such as the DES daughters will be among the most severe because of the difficulty in establishing negligence in the distant past. The ten year cut-off proposals, therefore, most prejudice those who would otherwise have been among the plaintiffs who stood most to gain from the introduction of a strict liability regime. Despite this, adoption of the cut-off period was recommended by both the Pearson Commission and the English Law Commission, the latter focusing principally on the injustice done to those injured in the latter stages of a product's life and not even mentioning latent damage.[29] Concern with the problem of latent diseases has simply not proved significant or at least sufficiently persuasive here.

Thirdly, the persistence of the neglect of disease issues is shown in the confusion which remains over the meaning of 'damage' in disease cases and this still clouds the important issue of apportionability. (Significantly, it is a question which has also been neglected by courts and reformers in the context of latent physical

[27] See *Pirelli General Cable Works Ltd.* v. *Oscar Faber & Partners* [1983] 2 AC 1.

[28] Council of Europe, *Convention of Products Liability in Regard to Personal Injury and Death* 27.1.77, Article 10; Council of the European Communities 'Council Directive on Liability for Reflective Products' 25.7.85, *Official Journal of the European Communities*, 7 Aug. 1985, L210/29.

[29] Pearson, para. 1268; *Report of the English and Scottish Law Commissions on Liability for Defective Products*, Law Com. 82 (1977) Cmnd. 6831, paras. 151–3.

damage.[30]) The fact that disease case-law fails to provide a clear statement of principle on such a central issue as what latent 'damage' is constituted by is indicative of the poor attention attracted to the application and adaption of traditional concepts to cases involving diseases of gradual contraction. To be fair, however, this lack of conceptual development owes as much to the paucity of disease case-law as to the neglect of advocates and judges. The following chapter, therefore, deals with the principal reason why so few man-made diseases result in the initiation and success of compensation claims: medical causation.

[30] See *Pirelli's* case *supra* and Law Reform Committee, *Twenty-Fourth Report: Latent Damage* (1984) Cmnd. 9390.

3

Causation in Fact

3.1 *Introduction*

The relaxation of limitation rules and other time constraints on obtaining compensation is of no value to disease plaintiffs who are unable to prove that the hazard under the defendant's control was the medical cause of that disease. Where, for example, a particular hazard produces genetic damage which does not manifest itself until a generation or more later, proof of causation will be impossible as a practical matter. Similarly, a causal nexus will usually be undiscoverable where the hazard has only subtle effects such as impairment of growth, intelligence, reproductive ability, resistance to disease, or behavioural abnormalities, or where the hazard produces only a marginal increase in an otherwise common disease. Proof of medical causation is also extremely difficult where the disease was not due to, say, the victim's direct use of a product or direct exposure to hazardous work conditions, but where the source was more generally or indirectly dispersed in the environment. For these and other reasons the aetiology of man-made disease presents the most complex conceptual and practical challenges to both the tort system and 'no fault' compensation schemes which are limited on the basis of cause of disability. This chapter discusses these difficulties and the more particular problem of establishing a causative relationship between the victim's disease and the specific hazard under the defendant's control which is the subject of complaint.

Throughout this discussion of how well the tort and other compensation schemes have, or could have, developed analytical frameworks to deal with them, the twin complications of multiple causation and human susceptibility recur, indicating the intrinsic complexity of the causation issue in the disease context.

3.2 *Establishing that the Hazard Can Cause the Disease*

In the tort system causation is seen as a fundamental and essential element of fault liability: a party will not be held liable for personal injuries unless his conduct actually caused the damage suffered by the victim. Thus, a manufacturer who carelessly exposes consumers to a carcinogenic product escapes liability if none contracts cancer. The first step in showing that the victim's disease was caused by the hazard under the defendant's control is to establish that the particular hazard *can* cause the disease. The discovery of medical causation-in-fact is, however, a slow and haphazard process, not least because the average general practitioner is more concerned with treatment than with aetiology. There often appears to be nothing to be gained from an investigation of causation, especially with conditions of low recognized incidence. As more cases are recorded, more interest may be expected to be shown firstly in the interests of prevention and secondly, as the number of treatment case histories increases, as a guide to the most suitable treatment. But, in general, the inexperience and inadequate training of the general practitioner is still widely acknowledged to be a major barrier to the early recognition of the origin of diseases, including those due to the activity of the medical profession itself such as adverse drug reactions. Yet even a well-trained and attentive general practitioner may be unable to discover the relevant causal connection because, for example, the class of people exposed to a particular dangerous product or work environment, is spread over a large number of doctors with the effect that the incidence of disease appears diluted to the individual doctor, particularly if there has been mobility of the affected group away from the hazardous conditions. This feature is exacerbated where symptoms do not appear until many years after exposure. What is more, latency periods vary between individuals so that, even when general estimates of latencies are available, they cannot be used in an individual case to determine (except in the crudest way) when and where the damaging exposure occurred. When the disease was contracted after a low dose exposure, the causal connection is, of course, even less likely to be detected. Inadequate exposure records are another factor which often exacerbates the problem. An illustration of these points is that it was only because workers exposed to vinyl chloride monomer presented latent symptoms of a particularly rare and fatal disease—angiosarcoma of the liver—that that causative link was discovered, significantly by a company doctor, while in the pharmaceutical field it was the rarity of the latent vaginal cancer caused by diethylstilbestrol that led to discovery of a causal nexus.[1]

Even when a causal link is suspected, a complex set of data, including the occupational, family, clinical, and social histories of the victims, may be needed before it can be accepted that the hazard causes this type of disease. Such information may be unavailable or at least difficult to obtain, for example, because the victim may not have been aware of his exposure to a hazard. In the work context, even the employer may be unaware of the often-changing composition of chemicals supplied under trade names and there may be difficulties in obtaining such information, especially in early speculative inquiries about medical causation. The likelihood of the victim being aware of past exposure is particularly low in cases of indirect exposure such as disease caused by industrial processes in which the individual was not directly involved, products used by another (as in the DES cases), and hazards to which large undefined classes of population are subjected such as pollution of the environment, food-chains, etc. Consequently medical causation is rarely establishable in such 'environmental' cases.

There may be misleading physiological indications: the disease may manifest itself, at least at first, by common symptoms such as headache, cough, or nausea, and the organs affected may not be those in direct contact with the hazard. For example, toluene di-isocyanate gives asthma-like symptoms and betanaphylamine causes cancer not in the skin and lungs where it is absorbed but in the organs (bladder and kidneys) involved in the excretion of its metabolites. Secondly, and more importantly, the disease itself may be contractable from other recognized sources which mask recognition of the alternative source. Even when the latter is suspected, a distinction between the two sources may be clinically difficult to make (e.g. some dermatitises) or impossible given current knowledge (e.g. many cancers). Epidemiological studies which in theory could achieve these distinctions may be impossible because of factors such as the smallness of the exposed group, the lack of data at chronic low level exposures, the unreliability of exposure records especially with long, latent diseases, or simply

[1] R. Livock, 'Science, Law and Safety Standards' (1979) 6 *Br. J. of Law and Soc.* 172; R. Vagley and M. Blanton, 'Aggregation of Claims' (1981) 16 *Forum*, 636, 639.

the prohibitive time and cost required to produce statistically significant data. Finally, as Ison has noted in the occupational disease context, there may be institutional barriers to the recognition of the origin of disease such as the priority given to traumatic injuries; the discouragement of the free flow of information by the ambivalent position of scientists employed by the party in control of the hazard as well as other health professionals; and the inability of victims or their representative bodies such as unions to fund independent research while industry can dictate the direction and dissemination of the results of research.[2] On the credit side, establishing a causative link may sometimes be assisted by a relatively close temporal or spatial nexus with exposure (for example, certain industrial dermatitises, scrotal cancer, etc.). Similarly, the investigations following the discovery that a substance causes a certain disease may lead to the discovery that it causes other ill effects, for example, the discovery that DES caused cancer in daughters of those who took the drug during pregnancy led to the discovery of DES-related low dose chronic diseases amongst the manufacturer's employees.[3]

Even where medically it has been established that a particular hazard, say a drug or an industrial dust, can cause the type of disease suffered by the victim, some disease victims face a further unusual barrier to discovering this fact. This factor, insignificant in accident cases where the medical causal link between the trauma and at least some injury is usually clear, is that of euphemistic or misleading medical advice and the withholding of associated information by concerned parties such as employers and social welfare officials in an attempt, for example, to save the victim from anxiety.[4]

[2] T. Ison, 'The Dimensions of Occupational Disease' (1978) 118 *Can. Med. Assoc. J.* 200.

[3] E. Shmunes and D. Burton, 'Urinary Monitoring for Diethylstilbestrol in Male Chemical Workers' (1981) 23 *J. Occup. Med.* 179.

[4] *Stokes* v. *Guest Keen and Nettlefold (Bolts and Nuts) Ltd.* [1968] 1 WLR 1776; J. McGuire, 'The Practitioner Looks at the Tough Occupational Disease Case' (1979) 14 *Forum*, 486–7 *Industrial Injuries Advisory Council Report: Pneumoconiosis and Byssinosis* (1972–3) Cmnd. 5443, para. 136; Medical Research Council, *Responsibility in the Use of Medical Information for Research* (London, MRC, 1972), 6; *Report of the Committee on Limitation of Actions in Cases of Personal Injury* (1962) Cmnd. 1829, para. 7; BMA, *The Handbook of Medical Ethics* (London, BMA, 1981), 13, 39, 86.

3.3 *Cause in Fact and Multiple Causation*

Once a victim knows that he has been exposed to a particular hazard and that that hazard has been linked to the disease from which he knows he suffers, he still faces the major problem of proving causation in his particular case. Until the exact pathogenesis of the disease is known the most an expert witness will be able to state is that on past observations certain persons exposed to these conditions have developed the relevant disease. By far the greatest difficulties in proving causation in the disease context concern 'multiple causation', and three distinct classes of events can be described as due to such 'multiple causation'. Each raises different issues for the law and must be distinguished. Firstly, there is the simultaneously 'multiple sufficient causes' case, well explored by academic lawyers: *A* and *B* simultaneously shoot *C*, who dies, although either wound would have been sufficient for the damage suffered. This is, not unexpectedly, a rare situation and has few direct parallels in non-traumatic personal injuries, principally because of the non-coincidence of exposures. Moreover, in most cases causes can be notionally distinguished because one can be said to accelerate the resultant injury in time or aggravate the end result, death being the clear exceptional case. Secondly, there are cases of 'multiple cumulative causes' giving rise to damage of an undifferentiated kind. There are many examples in disease cases where, for example, after exposure to one or more sources of dust a worker contracts a disease like silicosis due to the cumulative exposure. Because it is medically known that each individual intake of silica contributes to the overall damage suffered, the legal problem here is one of apportionment of damage between these known causes, not causation *per se*. Thirdly, and this is where the problem lies, there are cases where there are 'multiple *possible* causes' sufficient to produce the damage and it is not known which in fact did in this case. This is typified by diseases such as many cancers which may have been triggered by one or more exposures to a hazard or by exposure to a series of hazards capable of producing the disease, and the question for the law becomes, was the operative triggering in this case due to the relevant exposure? For example, was this victim's mesothelioma cancer triggered during his exposure to asbestos under a former employer or under his current employer? Similarly, since it is known that nickel workers contract lung cancer from their work, when one who also happens to be a smoker contracts that disease the question arises: was it triggered by the work or the smoking hazard (or both)? Finally there will be cases where not all the sources of risk of the disease which are under the defendant's control are the subject of the claim of negligence, in which case the victim must show that his condition was caused by the particular source of risk under the defendant's control which is the subject of complaint. In the case of multiple possible cause cases the causation issue must be settled as a question of relative probabilities between the possible causes and where—as is typically the case—these are unavailable, the problem is unsolvable under traditional theory. This latter class of case is enlarged by the large number of disease cases where the aetiology is inadequately known so that one cannot say that it is contracted gradually or by a triggering mechanism so one cannot in turn say that a certain exposure must have had some contribution to the damage, only that it might have.

3.4 *Proving Cause in Fact in an Individual Case: The Balance of Probability Test*

The plaintiff suing in negligence or for breach of statutory duty must prove a causal connection between his disease and the conduct of the defendant of which he complains. This task is eased by two legal principles; firstly, he need only show that the conduct was *a* cause even if only a minor contributory cause of his disease. In cases of multiple cumulative causes—such as deafness and the pneumoconioses—every exposure contributes to the condition of the victim. There is, therefore, no difficulty in establishing that some damage was contracted from the hazard and that it was a cause and not a mere condition. Consequently, in most cases of multiple cumulative cause there is no intrinsic difficulty in establishing medical causation—of at least some damage. Secondly, the victim need only establish a causal nexus on balance of probabilities. This helps in cases of multiple possible cause where the major problem is to show that the relevant hazard under the defendant's control, and not some other source of that risk, was the cause of the disease. If these alternative sources of risk were quantifiable by epidemiological studies and biostatistics a table such as Table 3.1 might be constructed. [*A* cases describe the

Table 3.1 Risk Analysis in Cases of Multiple Possible Causes

A Cases: Negligent Creation of Hazard *X*

	*A*1	*A*2	*A*3	*A*4
General risk	0.01	5.00	5.00	5.00
Total risk in presence of hazard *X*	1.00	5.02	10.00	materially greater than 5.00
Relative increase in risk due to hazard *X*	100	negligible	2	insufficient data
Chance that victim's disease is hazard-related	$\frac{100 - 1^*}{100} = 99\%$	negligible	50%	insufficient data
Balance of probabilities test . . . satisfied?	yes	no	no	insufficient data
Examples	rare cancer	bronchitis	lung cancer in nickel workers	lung cancer in nickel workers

B Cases: Negligent Omission of Safety Device

	*B*1	*B*2	*B*3
Risk with safety appliance	10	10	48
Risk with appliance omitted	30	materially greater than 10	60
Relative increase in risk due to omission	3	insufficient data	$\frac{60}{48} = 1.25$
Chance that victim's disease is due to omission	$\frac{3 - 1}{3} = 67\%$	insufficient data	$\frac{1.25 - 1}{1.25} = 20\%$
Balance of probabilities test . . . satisfied?	yes	insufficient data	no
Examples	silicosis	*McGhee's* case	Lord Salmon's example

* For explanation of formula, see T. Henderson, 'Medical Causation in Products Liability Disease Litigation' (1981) 17 *Trial*, 53, 57.

relevant comparison where the allegation is the negligent creation of hazard *X*: that is between the 'general' risk in the absence of hazard *X* (including sources of risk under the control of the defendant but not the subject of complaint) and the risk in the presence of the hazard. *B* cases concern cases of negligent

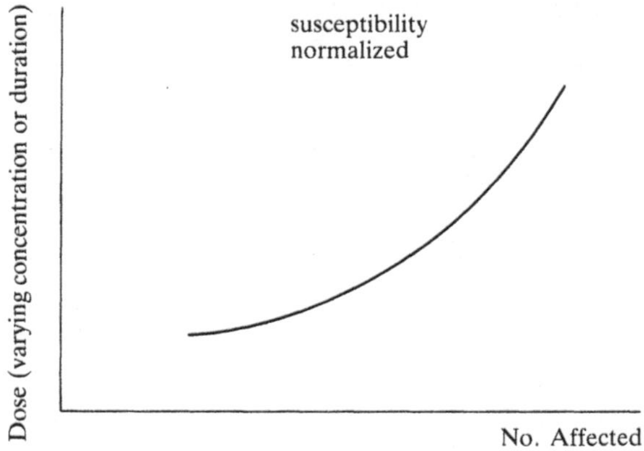

Fig. 3.1. Theoretical Dose-Response Curve

omissions: here the 'general' risk is the hypothetical risk level with the safety device deemed necessary in place and the risk level actually experienced, that is, with the device omitted.] Of course, risk depends on both individual susceptibility and dosage (length and concentration) of exposure—see Figures 3.1 and 3.2—so that even when available, such figures as listed in Table. 3.1 are at best crude averages over a whole range of factors, often bearing little relation to the actual risks encountered by the individual victim.

Let it be assumed, however, that these figures are available and sufficiently reliable. Take an allegation that the defendant was negligent in creating source of risk *X*: it must be shown that on balance of probabilities the victim's disease was due to the risk added by the relevant hazard under the defendant's control (that is *X*, the source complained of), and not to other sources of risk of the disease, that is, that the relevant hazard was the more probable cause (compare *A*1 with *A*2 or *A*3).[5] Mathematically, this requires that the chance of the disease being related to the relevant hazard is more than the chance of it being caused by an alternative source(s). The total risk in the presence of the hazard must be more than double that in its absence. A few examples will illustrate the point: suppose that in the general population 2 in 1,000

[5] In *McGhee* v. *National Coal Board* [1972] 3 All ER 1008, Lord Salmon appeared to overlook this requirement, stating that in situation *B*3 the 'employer could not escape liability' (1018).

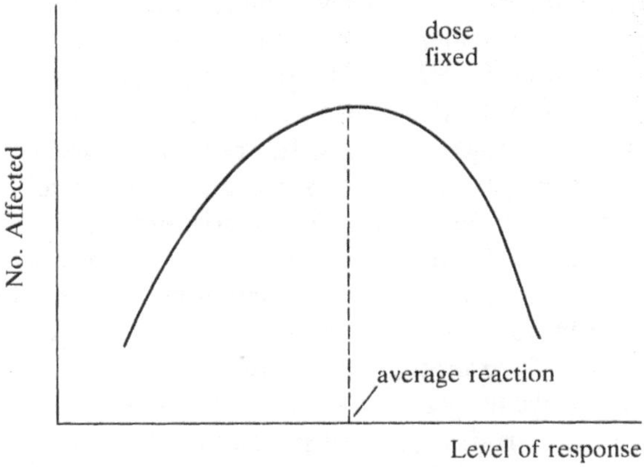

Fig. 3.2. Theoretical Susceptibility Curve

suffer bronchitis but in a certain industrial process *X* the incidence is higher, say, 3 in 1,000. Given a sufficiently wide data base it is possible to conclude that 1 in 3 victims employed in this industrial process is suffering a work-related illness, but there is no way to prove which one of the three it is. Overall statistics can alone be used. Moreover, only if the total incidence in the process is more than twice the general incidence (i.e. more than 4 per 1,000) can the individual victim show that his illness is more probably work-related than not. In situations where the disease's incidence is up to double or exactly double the general incidence, no individual can prove work-relatedness (so all fail to prove causation) even though, say in the latter case, half are in fact work-related cases. On the other hand, in situations where the workplace incidence is, say, 5 per 1,000, all victims of bronchitis from those workplaces can show that their illness was more probably work-related than not, even though two of the five are not work-related. Take another example: say a disease can be caused by a certain pharmaceutical and that product doubles the risk of that disease in consumers, then it still cannot be shown on the balance of probabilities that any individual case of that disease in consumers of the drug was due to the drug even though half such cases in fact are.

These examples and those in Table. 3.1 illustrate the intrinsic drawbacks to the all-or-nothing balance of probability test: the total risk with the relevant hazard present can be quite high—up to double that in its absence—yet will fail to satisfy the test so that all cases of the disease will go uncompensated even though half are in fact hazard-related (see *A*3). The problem is particularly acute in practice because most diseases are, or because of lack of medical knowledge on pathogenesis must be treated as, cases of multiple possible cause and reduced to this precarious balancing of crudely calculated probabilities. Moreover, in most of these cases even the minimal (albeit sufficient in law) assessments as one source being 'more probable than not' are not yet scientifically possible. Even in the case of fatal diseases, for example, the rarity of autopsies and the often vague, inaccurate, or misleading form of death certificates undermine the accuracy of any retrospective epidemiological study and so even the crudest relative probabilities are unavailable. In such cases it is simply not possible to satisfy the traditional balance of probabilities test and the victim fails to win compensation. The overall result is that, while most accident victims have no difficulty in showing a causal connection between the wrongdoer's hazard and his damage, this is a major and characteristic barrier in most cases involving man-made disease. Because of the haphazard and inherently slow process by which evidence of causation is usually accumulated, and the prohibitive cost of independent research, early victims of such diseases, therefore, have virtually no chance of recovering common law damages, while even in claims by later victims the case will usually be very costly to mount and the outcome may—in an adversarial system—depend heavily on the credibility, rather than the technical ability, of experts who may genuinely hold greatly differing views on the medical evidence.

3.5 *Causal Link to the Subject of Complaint*

There is a further complicating factor in proving causation in the typical disease case which was mentioned briefly earlier. This is that it is not enough for the victim to prove that his conduct was due to a hazard under the defendant's control but that (except in a few cases where a claim of strict liability is available, for which see Chapter 4) it was caused by a source of risk which was itself due to the negligence of the defendant. For example, say that a plaintiff argues that his condition was due to an adverse reaction to a pharmaceutical product marketed by the defendant. Say that the basis of the legal complaint is an allegation that the packet warnings on the product were inadequate; if there were batches of the drug released by the defendant with adequate warnings and batches released without these, then the plaintiff must show that his condition was due to a product in the latter batches. A clearer and more common example from the reported case-law concerns dust diseases. Say the plaintiff has contracted an asbestos-related cancer after working in an asbestos mill and say there were sources of asbestos dust which were 'innocent'—that is, not the subject of the complaint of negligence—and sources which are 'guilty'—that is, the subject of complaint. Then once again the victim must prove on the balance of probabilities not simply that his cancer was work-related but that it was due to the source of risk of which the complaint of negligence is made. In other words, this means that even if a disease is only contractable, say, from work conditions (byssinosis and silicosis are common examples), and even if one assumes that a disease victim had only ever worked for the defendant, the plaintiff still has a causation hurdle to jump: he must show on balance of probabilities that it was those work conditions which are the subject of the complaint which were a cause in fact of the disease. In cumulative diseases such as silicosis, there is no problem in showing this because as long as exposure due to these 'faulty' conditions was material, it will have caused at least some material damage (although assessment of relative risks will still be critical to the causal apportionment of damage). There is a problem, however, in cases of multiple *possible* causes such as dermatitis, cancers, etc. because the disease may have been triggered by conditions which are 'innocent' (that is, not the subject of complaint). Here the plaintiff must show that it was exposure to the faulty conditions which was the more probable cause of triggering. So even in cases where the only possible source of this disease was under the defendant's control, the vagaries of the all-or-nothing balance of probability test just outlined will still be involved.

Moreover, at this stage an added complication must be noted: that is, the faulty conditions may be attributed to an omission, say, the omission of warnings on a product label or of safety equipment in a workplace. To appreciate the additional problems of causal proof this poses it is necessary to move from purely scientific inquiries to a consideration of how events are described. The law distinguishes acts of commission and acts of omission, a distinction which springs from liability rules—how and by whom the law expects the hazard to be controlled—rather than a scientific description of physical reality. The worker exposed to an industrial dust or carcinogen may be able to argue that his employer was at fault in exposing him to it at all (an act of commission), or alternatively that, the exposure being reasonable, the lack of adequate safety measures such as ventilation or masks was wrongful (an omission). Where the plaintiff succeeds in the argument that it was wrongful *per se* to expose him to the hazard, causation in fact is sufficiently proved by scientific evidence that exposure to this guilty source of risk contributed to a cumulative disease or was the more probable cause of triggering a disease of multiple possible causes. However, for reasons explored in Chapter 4 the disease plaintiff usually relies on the wrongfulness of an omission—for example the omission to provide adequate safety equipment such as masks, or adequate product warnings and directions for use. In such cases the burden which under traditional rules falls on the plaintiff involves much greater difficulty: to show that, on the balance of probabilities, the victim's disease was due to the omission. To do so it must be shown that: (i) the victim would probably have used the omitted device or followed the product advice; and (ii) the omission of the device had more than doubled the likelihood of the disease (or exacerbation) which is the subject of the complaint.

Again, the fundamental problem is that the plaintiff can only deal in probabilities, but with omissions these are clearly much more difficult to assess. Firstly, there is an intrinsic difficulty in hypothesizing what a person's probable behaviour might have been in an imaginary situation, and some hard cases testify to this.[6] Because of these difficulties culpable defendants might escape liability even though they have been in flagrant breach of duty. For policy reasons this is an unattractive result and of particular importance in the disease context where cases almost invariably proceed on the basis of an allegedly culpable omission. One solution might be to shift the onus of proof on the causation issue to the defendant. On the issue of whether the victim would have used a safety device if it had been provided this has not been done, which perhaps reflects the difficulty in overcoming conventional theory, or at least the relatively less onerous problem of proof on

[6] See e.g. *McWilliams* v. *Sir William Arrol & Co.* [1962] 1 All ER 623 (accident case); *Richards* v. *Highway Information Ltd.* [1957] 2 All ER 162 (disease case).

the balance of probabilities with respect to human behaviour as compared with issues of medical causation (point (ii), see below). At least with human behaviour there will be evidence of character, habit, etc. from which to estimate, if only crudely, what the person would have done. For example in a simple case the court may be willing to assume that had the defendant provided the wrongfully omitted medical monitoring of the victim, the latter would have availed himself of it.[7] In other cases where the safety precaution would have involved greater discomfort or inconvenience for the plaintiff—such as wearing a mask constantly while at work—the plaintiff's task in establishing (i) is more complex and difficult. Some courts have lightened the plaintiff's burden here, however, by imposing in relation to the issue of a hypothetical safety device a duty on the defendant to instruct, exhort, and perhaps supervise. The plaintiff then has the less onerous burden of showing that had such instruction, etc. occurred he would have used the device.[8] None the less, the conventional treatment of the plaintiff's conduct as a matter of all-or-nothing causation continues to pre-empt a more attractive analysis based on relative blame (contributory negligence)—much to the disadvantage of many victims (see 4.10).

The second causal barrier in omission allegations is the difficulty in assessing the probable efficiency of the omitted safety precaution in the precise conditions in issue (see (ii) above). The problem is one of multiple possible causes: the two risks being compared being the general (acceptable) risk with the hypothetical safety precaution deemed necessary in place and the risk level actually experienced in its absence. This is illustrated by the *B* cases in Table 3.1. In the context of omissions the proof problem is clearly even more difficult for the victim because not only does he have to prove issue (i), but this comparison of probable risks in issue (ii) involves an expert imagining a hypothetical precaution in place and estimating the resultant reduction in risk. In short, in such cases there is even less likelihood than in other multiple possible cause cases that experts will feel confident about giving the necessary estimate to support the victim's case on causation.[9]

[7] *Wright* v. *Dunlop Rubber Co. Ltd.* (1972) 13 KIR 255, 256; *Stokes* v. *Guest Keen & Nettlefold (Bolts & Nuts) Ltd.* [1968] 1 WLR 1776, 1797.

[8] See e.g. the disease cases *Bux* v. *Slough Metals Ltd.* [1974] 1 All ER 262; *Berry* v. *Stone Manganese & Marine Ltd.* [1972] 1 Ll. LR 182, 187. See also *Nolan* v. *Dental Manufacturing Co. Ltd.* [1958] 2 All ER 449 (accident case).

[9] For cases where it was held that the omitted safety precaution would have been

In dealing with this problem a most radical departure from established doctrine was employed by the House of Lords in *McGhee*[10] which concerned what was indisputably a disease caused by the pursuer's work conditions. The defendant-employer admitted the negligent omission of a safety precaution (washing facilities) but denied the pursuer's allegation that it was this source of risk which had caused his dermatitis condition. The issue was, therefore, what was the efficiency of that omitted precaution. In short, all the pursuer could show was that the negligent omission had materially increased the *risk* of his contracting the disease. Medical experts were unable to estimate the relative probable contributions of the two sources of risk—the general risk and that in the absence of the washing facilities. The result was that the pursuer could not show that the risk added by the negligent omission had in fact contributed to his disease, nor that it was, on the balance of probabilities, the more probable cause of it or of its aggravation than the background or 'innocent' source of risk (that is, that which would remain even in the presence of washing facilities). So a traditional burden of the preponderance of probabilities could not be discharged on the issue of medical causation (see Table 3.1, *B*2).

Despite this, the Law Lords held that the pursuer had sufficiently proven a causal link between his disease and the omission complained of. This finding, that a material increase in risk is sufficient to prove causation, has gone undeveloped by later English case-law but constitutes a remarkable rejection of the traditional preponderance of probabilities concept as inadequate when dealing with the classic multiple causation problem in disease cases: proving that a negligent omission was the medical cause of the disease where estimates of relative risks are unavailable. In Lord Wilberforce's opinion the inherent evidentiary difficulties in cases like *McGhee* should as a matter of policy be borne by the party who has created, or materially increased, the risk faced by the victim.[11] Whatever the merits of this approach in cases where all sources of risk were under the control of the defendant, in the more common situation, where other possible

ineffective see *Brown* v. *Rolls Royce Ltd.* [1960] 1 All ER 577 (disease); *Robinson* v. *Post Office* [1974] 1 WLR 1176 (trauma).

[10] *McGhee* v. *National Coal Board* [1972] 3 All ER 1008.

[11] Ibid., 1012. The same policy underlies the market-share doctrine.

sources of risk are outside the defendant's control, considerations of deterrence and fairness to the defendant arise more strongly. This is where the reasoning in *McGhee* is inadequate. In economic deterrence terms the Lords' 'solution' to the victim's causation dilemma overinternalizes losses to the defendant to an even greater degree than the traditional regime of an all-or-nothing balance of probability test (which overinternalizes to the extent that it fails to reflect causal contributions such as a substandard road surface in road accidents). There is little, if anything, in the *McGhee* reasoning to prevent it being used by a person who can show that his unreasonably (that is, negligently) stressful work conditions materially increased his risk of heart attack or disease. Why would he not be able to use the reasoning of the Law Lords to establish a causal link to his heart condition even though its non-occupational causes were clear and far more significant? Why should the employer bear the cost of the heart condition?

In theory, overinternalization might be reduced by an accompanying rule of reduction of damages recovered to take account of possible sources of risk other than that which is the subject of complaint. Causal apportionment like this is feasible across the whole range of contribution to risk and allows a more accurate internalization of costs. It would assist, for example, in multiple cause cases which currently fail—and consequently *underinternalize* costs—because the risk attributable to the guilty source is estimated at less than 50 per cent. In the United States the market-share doctrine is an example of these dual goals of a more refined allocation of losses, and therefore deterrence, coupled with allowing recovery in cases of multiple possible sources of risk—in this case where medical causation is established to a generic product but there are multiple possible tortfeasors, none of which can be identified on the balance of probabilities to have been the more probable source of the product in the individual case.[12] Causal apportionment is also a valuable technique in incorporating a more sophisticated level of deterrence in the funding of a no-fault compensation scheme (see 6.8). But the

[12] Apportionment is made on the basis of market share as this reflects the increase in overall risk to consumers due to the defendant's product. See G. Robinson, 'Multiple Causation in Tort Law: Reflections on the DES Cases' (1982) 68 *Virg. L. Rev.* 713. A proponent of apportionment in the wider disease context is D. Rosenberg, 'The Causal Connection in Mass Exposure Cases: A "Public Law" Vision of the Tort System' (1984) 97 *Harv. L. R.* 849.

conceptual refinement of causal apportionment on the issue of medical causation was not mooted in the *McGhee* reasoning, nor in practice would it be of much, if any, assistance to disease victims, because it hinges on the very estimates of relative contributions to risk which are typically unavailable in such cases. In other words, the device which would reduce the distortion and anomalies produced by the Law Lords' 'solution' to the causation problems of the disease victim would fail in those very cases where the problem is at its most intractable. In terms of conceptual development to accommodate the disease victims' problem it is a dead end.

To sum up: the approach of the Law Lords in *McGhee* to the typical multiple possible causation problem in a disease case shows how extreme the departures from conventional theory must be before any attempt can be made to accommodate these cases. The result is the creation of anomalies and a distortion of legal rules which, in any case, ultimately seems to prove a dead end. This might be an explanation of the relative neglect by academics in this country of such a dramatic rejection of a fundamental doctrine as the balance of probability test,[13] although one suspects that it is also yet another good illustration of the poor appreciation of the important challenges to doctrine that disease issues have presented. On the other hand, the failure to address the doctrine in disease case-law is probably better explained by more practical factors such as the relatively few disease cases reaching the courts and the fact that, of these, very few concern areas of aetiological doubt. In short, it would seem that the very factors which would reveal the common law's analytical framework to be inadequate—here the causation dilemmas of the disease plaintiff—are those which deter litigation.

A final implication from the approach taken in the *McGhee* decision: here we saw the Law Lords, faced with the typical causation barrier to disease compensation, abandoning a traditional element of the one-to-one corrective justice model, namely that the defendant be proved to be more probably than not the party who caused the plaintiff's damage, and allowing recovery against a

[13] There has been more interest by academics from North America: G. Robertson, 'Overcoming the Causation Hurdle in Informed Consent Cases: The Principle in *McGhee* v. *N.C.B.*' (1984) 2 *Univ. of W. Ontario L. R.* 75; J. Fraser and D. Howarth, 'More Concern for Cause' (1984) 4 *Legal Studies*, 131.

defendant who had merely been shown to have negligently added one of a number of possible sources of risk. If this is the best solution to the problem which the common law can provide, it suggests that an accommodation of disease issues would weaken the traditional basis of tort liability. As Calabresi has noted,[14] once it is accepted that justice does not require an individual wrongdoer to compensate his *actual* individual victim it becomes more difficult to support the fault system in terms of justice. If the interrelationship of larger groups of people than a specific plaintiff and the defendant who *actually* caused his injury are to be taken account of—and this is what *McGhee* implicitly does by expanding the class of potential defendants to include those who can merely be shown to have increased the risk of injury—then the focus shifts towards the sort of distributive justice focus reflected in non-tort compensation funds, where payment of compensation is not predicated on the identification of actual injurers and compensation is detached from the collection of contributions (which are in turn made by those risk creators who can be identified). Not only do the practical problems of proving disease causation suggest abandonment of tort in favour of a compensation fund, but this appears to be the more rational way to accommodate policy goals such as fairness, compensation, and even deterrence (see Chapter 6). An historical parallel may be drawn here: just as workers' compensation arose as a response to the inadequacies of the tort remedy in practice in the area of work accidents, so too the most rational solution to the inexorable causation problems now being recognized as confronting disease victims seems to be replacement of tort by a non-tort system. The question then arises: how well do non-tort compensation funds handle the problem of disease causation?

3.6 *Causation and No-Fault Compensation Schemes: The Accident/Process Demarcation*

Under no-fault compensation schemes causation questions are both simplified by elimination of the requirement of an identifiable wrongdoer and complicated by broader issues of policy as to how a limited compensation scheme can be run fairly and rationally. The principal examples of schemes covering disease are those which

[14] G. Calabresi, *The Costs of Accidents* (New Haven, Yale Univ. Press, 1970), 302–3.

only cover disease of an occupational origin such as workers' compensation and the industrial injuries schemes. These are defined and limited by the causal requirement that the disability be caused by an 'injury by accident arising out of and in the course of employment'.

This critical phrase originated in the Workmen's Compensation Act 1897 (60 and 61 Vict., c. 37) which set up a scheme initially intended to cover only traumatic work injuries, not sickness. [For diseases a system of sickness insurance where cause would be irrelevant had—prophetically—been considered more appropriate, but it was not proceeded with.[15]] It was, therefore, something of a surprise when the House of Lords in 1905 held that infection by a bacillus of anthrax was a sufficient incident to render the victim's disease an 'injury by accident' and compensatable under the scheme.[16] That the 'accident' might be microscopic and intrinsically unobservable probably still remains somewhat outside the common sense meaning of 'accident'. But in practice the extension of coverage to diseases of infection was more understandable since the disease would have been compensatable had it been traceable to an overt trauma such as a cut, no matter how minor that cut had been. Had the infiltration of diseases into the accident compensation scheme been halted at this stage, the integrity of the coverage rule might have been adequately preserved—at least as a formal demarcation line between traumatically caused injuries and ill health. But in a series of cases on disease claims following the anthrax case the courts developed the notion of an accident consisting of a series of incidents. Once this step had been taken—arguably on even greater departure from common usage than the 'unobservable' accident notion in the infection cases—the logic of the distinction being made disappeared. Rationally there is no meaningful distinction to be made between diseases of even the most gradual contraction process and those attributable to 'a series of incidents'. Medical usage itself makes no sharp distinction between injury by accident and disease. Nuances of meaning were, however, developed. The rapidity and regularity of a pathogenic attack or stressful event such as muscular stress, for example, were fixed upon as factors militating against accident and in favour of it

[15] *Departmental Committee Report on Workmen's Compensation* (1904) Cd. 2208, paras. 104–5.

[16] *Brinton's Ld.* v. *Turvey* [1905] AC 230.

being a 'process' and as such non-compensatable. Thus skin diseases caused gradually by long exposure to fumes or chemicals were not covered, but the same disease precipitated by a day's exposure to chemicals was. There was no change in the 'injury by accident' coverage formula when the Industrial Injuries Scheme replaced workmen's compensation in 1948, and the anomalies clustered about the accident/process distinction have, therefore, survived: a claimant who had suddenly felt severe pain on lifting a heavy object and later suffered disability from coronary thrombosis was covered, but muscular strain caused by regular lifting was not.[17] Similarly, the New Zealand scheme set up in 1972 covered cases of 'personal injury by accident', and in interpreting the phrase the Accident Compensation Commission appears to be developing a similar 'process' distinction to that used in the limited employment schemes.[18]

The conceptual irrationality of the distinction is now generally accepted. The many cases where there is doubt about how a disease is contracted—whether by an isolated event or cumulative effects—highlights the potential arbitrariness of the rule in practice. But while modern critics are correct to emphasize the injustice of the distinction in such cases, it is wrong to see the uncompensated 'victims of the doctrine of process' diseases as anomalously excluded from coverage. In the context of original Parliamentary intent, the anomalies are those suffering from other diseases (such as those of infection) who are nonetheless able to bring themselves within the general formula of 'accident'. It is arguable that within the common meaning of the term 'accident' these fortunate claimants are still anomalies today. Moreover, the distaste held by some reformers for the 'process' rule has resulted in a somewhat distorted emphasis being placed on its role as a barrier to recovery of compensation for disease victims. It is in fact of minor importance. Had the 'process' debate been pre-empted by the use in the coverage clause of a term such as 'disability', disease claimants would still have faced the fundamental problem: that in cases not involving a visible trauma isolated in time and space there are intrinsic—often insuperable—difficulties in establishing work-relatedness.

[17] R(I) 54/53 and R(I) 42/51 respectively.

[18] See G. Palmer, *Compensation for Incapacity* (Wellington, OUP, 1979), 251; e.g. *Re Rivers: Decision 773* [1982] NZACR 321.

3.7 *Scheduling Guide-lines and Systems of Individual Proof: The Intractable Problem*

Once anthrax had been held to be compensatable under the Workmen's Compensation Scheme, concern was raised when in later decisions compensation was refused for diseases of gradual contraction (or 'process'), such as lead poisoning and beat hand or knee, which were widely recognized to be just as specifically work-related as infectious diseases such as anthrax.[19] The result was the statutory extension of compensation to victims of certain diseases in certain occupations listed in a schedule to the Workmen's Compensation Act 1906 (6 Edw. 7, c. 58). This development in the 1906 Act was not seen as the first step to the comprehensive coverage of all occupational diseases, but as the first of a series of *ad hoc* extensions to only those diseases which could be shown—to a high level of proof—to be related to specific industrial processes.

Until recently this has continued to be the dominant philosophy governing the scheduling of occupational diseases under the workmen's compensation and later the industrial injuries schemes. Under this view, to schedule a disease on the basis merely that overall such disease was more likely to be work-related than due to other causes would be unfair. It was thought unjust to claimants that compensation should depend on such an all-or-nothing rule—especially, as typically the case, there were inadequate statistics—but more importantly it was unsatisfactory that the employer (under workmen's compensation) or the industrial injuries fund be called upon to pay all claims for a disease including the indistinguishable percentage which was not work-related. At first this approach led to a requirement that contraction of the disease due to non-work factors only occur 'very rarely'.[20] But the major disadvantage of this approach to scheduling was to exclude any disease which had a significant incidence among the general population even if its incidence due to work was much higher (such as tuberculosis among health workers). Subsequently, the policy relaxed to allow the scheduling of common diseases but only where the incidence was much higher in the relevant occupation *and* where work-related attribution in the individual

[19] *Steel* v. *Cammell, Laird & Co. Ltd.* [1905] 2 KB 232; *Marshall* v. *East Holywell Coal Co. Ltd.* (1905) 7 WCC 19.

[20] *Departmental Committee Report on Industrial Diseases* (1907) Cd. 3495, p. 3.

case was possible 'with reasonable certainty'.[21] In such cases common diseases might be covered although only by way of 'open prescription', that is, prescription without the usual access to a rebuttable presumption of work-relatedness. [The stringency of qualifying conditions of prescription such as exposure or narrow occupational limitations could also be used to increase the statistical probability of work-relatedness to the required level in some common diseases.] The paradigm example of this relaxed presumption is industrial dermatitis. In the case of other common diseases such as bronchitis, where such a clinical distinction could not be made in the individual case, the fund and adminstrative resources would be drained by claims not intended to be covered by the scheme. In such cases, prescription was regarded as inappropriate.

Critics of this view argue that it is too conservative, that if a disease is more likely to be work-related than not it ought to be scheduled. The difference between the two views is simply one of the degree to which the fund can be opened up to non-occupational claims without its preferential basis being seriously eroded. A similar tension can be seen at work in recent moves towards a system of individual proof which would exist alongside the cautious and consequently slow scheduling system.[22] [This system would allow a victim of an unscheduled disease the opportunity to prove it was work-related in his case. The idea is, at least superficially, appealing because there are many poignant cases—some involving fatalities—where there is no dispute that the victim's illness was due to work conditions but the claim fails because the case cannot be classified as an 'accident', nor is it a disease for which a general scheduling formula can be designed because the hazardous circumstances are not general to the victim's working conditions but specific to his own work experience.[23]] A number of devices have been suggested by which

[21] *Departmental Committee Report to Review the Disease Provisions of the National Insurance (Industrial Injuries) Act* (1955) Cmd. 9548, majority report, paras. 86–8. Cf. minority report, para. 8.

[22] See e.g. *Departmental Committee Report: Industrial Diseases: A Review of the Schedule and the Question of Individual Proof* (1981) Cmnd. 8393 (hereafter the 'Lowe Report'), and comments thereon by R. Lewis, 'Compensation for Occupational Disease' [1983] JSWL 10; S. Wilson, 'Occupational Disease: The Problems of a Comprehensive System of Coverage' (1982) 1 *Ind. L. J.* 141.

[23] See e.g. R(I) 7/66, R(I) 4/80, and examples from jurisdictions where industrial compensation is not limited by a schedule: *Re Bishop and Commonwealth of*

non-occupational claims, especially involving common diseases, could be excluded from this system, but government has so far—apparently on the grounds of avoiding the costliness of overloading the system with cases doomed to fail for lack of medical evidence—rejected suggestions of an individual proof system, even one qualified with such devices.[24] But the essential dilemma continues to be neglected by reformers, that is, that the more generous the scheduling guide-lines and the more available is the device of individual proof, the more non-occupational claims will be admitted to the scheme, calling into doubt its preferential basis itself. In other words, in the typical case where the aetiology of a particular disease can only be 'proven' by statistics, there is an inevitable clash between the twin goals of a coverage rule of a scheme limited on the basis of cause: to include those cases of disease which fall within the preference and to exclude those which do not.

Just as this problem arises if priority is given to the exclusion goal (by a cautious coverage rule), so too it persists where the focus is on including all valid claims. This focus may arise for different reasons such as the special position of the armed forces or the preference often encountered in statutory schemes for those suffering work-related injuries.[25] But whatever the reason for an inclusionary emphasis the coverage dilemma will still arise. For example, assume the goal is to compensate work-related leukaemia. Having examined the relative incidences of leukaemia among the general public and among nuclear industry workers in the United States, Estep concluded that the only acceptable solution by which those suffering work-related leukaemia could all be compensated and compensated fully was to set up an industry-wide contingency injury fund, into which risk-creating employers would make a

Australia (1982) 5 ALD 247 (employee winning compensation for the aggravation of his allergy by the smoking of his fellow-workers).

[24] See limiting devices suggested by Pearson, paras. 884–7; the Lowe Report, para. 159. But see rejection by government: HC Deb. vol. 52, written answers, col. 329, 20 Jan. 1984.

[25] See e.g. the favourable causation tests discussed in *Repatriation Commission* v. *Law* (1981) 36 ALR 411 (see 5.8); and in the US Black Lung Benefits programme which resulted in it being seen as 'now more akin to a pension program' by J. O'Connell, 'Foreclosing Claims for Personal Injury from Toxic Substances by Defendant's Tender of Claimant's Net Economic Losses' (1982) 2 *Virg. J. Nat. Res. L.* 203, 204 n. 8.

contribution for the increased risk due to work, and from which all nuclear workers who contracted the disease could claim.[26] Those whose disease was in fact due to natural incidence would therefore also be compensated—there having been a contribution into the fund to cover these cases by, preferably, a work-force-wide compulsory insurance scheme. That is, workers would insure against the low risk of natural incidence but be covered (by the industry contributions) for the total risk of the disease: in effect a compulsory health insurance plan for specific workers for a specific disease. The impossibility of proving causation in the individual case would then be rendered irrelevant. The scheme has the clear advantage of *fully* compensating the victims of this occupational disease which, for example, could not be accomplished by a solution based on causal apportionment. It spreads the cost of the disease in a fairly effective and equitable manner and covers all increase in disease risk due to work, even if less than the 'doubling dose' necessary under the traditional balance of probabilities test. The drawbacks are twofold. Firstly, statistics on incidences are in fact rarely available. Estep himself notes the unreliability of such data for other radiation injuries such as genetic defects, sterility, premature ageing, and increased susceptibility to disease in general. Even with leukaemia, which had the most reliable data, a 'guess' had to be made within a range of figures about what dose level would double the risk of the disease. The *McGhee* situation where medical science can, at best, say that work had 'contributed' to the risk of disease remains the more characteristic scenario for diseases. Secondly, there may well be policy objections to a situation where all victims of this disease among the exposed workforce recover, including those due to natural incidence—while similar natural incidence victims in the general population go uncompensated. If, as logically it should, the scheme were to be applied to *all* diseases to which work makes even a small contribution of risk, why should the advantage of such a general sickness insurance scheme be in fact limited (*a*) to these diseases, and (*b*) to employees? Rationally, the only way to provide compensation for all occupational disease is to abandon the requirement of work-relatedness itself as done in the Netherlands (see 7.7, Option IV). But once this is done, it is hard, then, to see why equal treatment is not in turn accorded to non-employees.

[26] S. Estep, 'Radiation Injuries and Statistics' (1960) 59 *Mich. L. Rev.* 259.

3.8 *Causation Problems in the Individual Case*

Even when a disease—say *X*—is one of the few which have achieved scheduling, this by no means resolves all the causation problems for the individual claimant. Not only does that victim have to survive all the threshold difficulties inherent in discovering that he has a disease which is linked to his occupation (see 3.2), but he must satisfy the authorities that it is the scheduled disease *X*. In practice this can often be extremely difficult because symptoms may be similar to those of non-scheduled diseases, especially at an early stage. In many cases the rejection of claims is found to have been wrong only upon post-mortem examination. Another difficulty—specific to occupational deafness claims—is that the victim will have to satisfy certain minimum and recent exposure conditions before he qualifies for benefit, conditions imposed in part to ensure diagnostic certainty (and in part to limit the volume of claims). Thirdly, the victim must not only show that he suffers from the scheduled disease *X* but that it was due to employment in prescribed conditions. When a disease is scheduled it is scheduled only in relation to workers in specific occupations or conditions, and the claimant must prove firstly that he was employed in such conditions and secondly that the disease was due to those conditions. Usually most workers in prescribed conditions will have access to a presumption that their disease was due to those conditions but there are some cases which fall outside the wide coverage of the presumption or concern one of the few diseases for which no presumption is available. In these cases, proof of the relevant causal nexus can be difficult because the intrinsic problems of the traditional balance of probabilities test persist and limit even further the number of claims which can successfully be brought within the effective coverage of the schedule. For example, just as at common law, there are cases where all that medical science can reliably testify to is that his work increased the risk of the victim's disease but cannot quantify this effect, with the result that there is insufficient evidence to show that the disease was *due* to that employment (for which, it must be remembered, there must be evidence that the employment more than doubled the risk).[27]

An interesting parallel between the situation at common law

[27] R(I) 33/51. Cf. *McGhee's* case.

and under the industrial compensation system can be noted here. Just as in *McGhee* the House of Lords attempted to bypass the inadequacies of the traditional causation test in the disease context, so too within the industrial compensation scheme there has been an unacknowledged *ad hoc* attempt to do the same in the context of one particular class of disease: heart-related incidents which can be classed as 'injury by accident'. Here we find that claims are allowed merely on the ground that the work conditions had helped in a material degree to bring about the incident, for example a heart attack or stroke.[28] The requirement that it was, on the balance of probability, more likely that the incident was due to the added risk from work is ignored. Again, in theory, the admission of such claims might be made more acceptable if the benefit was reduced to reflect only that amount of aggravation of the underlying heart condition shown to be due to the employment.[29] But again, in practice, it is these very estimates which are not available in reliable form. The result is an anomalous class of claims which succeed despite failure to satisfy the classic all-or-nothing causation test. Once again, the attempt to ameliorate the problem of the victim of a disease which has multiple causes, and for which adequate statistical estimates are unavailable, has led to an objectionable state of affairs. It is objectionable both at a policy level for its anomalous pattern of recovery—further compromising the industrial preference basis of the coverage rule by producing unacknowledged general insurance cover in the heart incident area—and at a more specific economic level for its crude overinternalization of costs, distorting whatever deterrent impact the system would otherwise have.

There is a final point to be made about the requirement that the victim of a scheduled disease must prove the disease was due to employment in prescribed conditions. It is a requirement inserted in the scheme to narrow the pool of claimants to a subclass of victims of the specific disease where it can be said, without investigation of the particular facts of the situation and to a high level of certainty, that the victim's disease was due to the

[28] *Clover, Clayton & Co.* v. *Hughes* (1910) 3 BWCC 275; R(I) 3/56; R(I) 14/51; R(I) 12/68; R(I) 42/53; R(I) 20/51.

[29] There is elaborate machinery to allow causal apportionment: Social Security (General Benefit) Regulations 1982, SI 1982 No. 1408, Reg. 11, but it appears to be unused in the heart cases.

conditions in which he worked. However, just as the lack of a system of individual proof excludes workers who are on the particular facts of their case indisputably suffering from an occupational disease, albeit unscheduled, so too the requirement of employment in prescribed conditions excludes cases of disease (both scheduled and unscheduled) caused by workplace hazards where the victim was not so employed even though on the facts of the case the disease is clearly attributable to such a hazard. This means that not only are, say, workers in an associated process or activity in the same plant excluded—such as the secretary in a plant handling asbestos who contracts mesothelioma—but more importantly it excludes all those non-workers who contract a disease (even a scheduled disease) which is due to a workplace hazard but who contract it indirectly via some environmental agency. Cohabitees of asbestos workers, and those who live nearby to an asbestos plant, can contract mesothelioma cancer from the asbestos dust with which they come into contact. Similarly, people can contract diseases attributable to industrial hazards via industrial effluents, waste storage sites, and other forms of environmental contamination such as agricultural chemicals drifting in air or seeping onto nearby land. In most cases the causal link to the industrial hazard will not be suspected or diagnosed with certainty, but in some cases—such as the mesothelioma cases—the nexus can be asserted with as high a degree of certainty as that required for the scheduling of a disease. But these people are not covered by the scheme even though they are just as much the innocent victims of the industrial process as those inside the factory gate who qualify. While the workers, who suffer from what is indisputably an occupational disease and yet are excluded, are victims of the exclusionary goal of a severely applied causal coverage rule, these environmental victims of industrial hazards are the victims of the wider and much less coherent dual policy that firstly, victims of the industrial process are to be given preferential treatment and secondly, the preference is only to apply to that subclass who contract the disease in the course of their employment (see 5.8).

3.9 *Conclusion*

This chapter has looked at the issue of proof of medical causation in disease cases. It ignored complex issues such as synergism—that is, where one hazard under the defendant's control combines with a natural factor (for example, carbon monoxide and lead; amphetamines and overcrowding) or man-made factor (for example, asbestos and tobacco; barbiturates and alcohol) to produce a multiplicative effect.[30] Even so the discussion has shown that the complexities of disease pathogenesis produce anomalies and injustice under both common law and industrial injuries systems. In particular, and unlike most trauma cases, medical causation in disease cases is typically a matter of abstract probabilities between multiple causes, at best only crudely assessable. This feature compounds the problems inherent in the all-or-nothing balance of probabilities test, which either under- or overcompensates classes of disease victim and in limited non-tort schemes calls into question any causally based coverage rule. Over all this produces even more of a 'lottery' for claimants than exists in trauma cases. In theory, a more rational solution might be found in both systems by the replacement of this test by a system of apportionment of damages between possible causes. However, under both schemes such a solution, requiring the collection of complex medical evidence, would be very costly, and, in any case, limited to those very few cases where a causal link is suspected, contributors to risk identifiable, and some assessment of relative contribution to risk—even if only crude—can be made. In the vast majority of man-made disease situations this will be impossible (3.1–3.2). The tort system simply cannot accommodate them, nor can any non-tort scheme limited on the basis of cause.

[30] How the law will deal with synergism is unclear, but asbestos manufacturers in the US have filed cross claims against tobacco manufacturers on this basis, and proposals for non-tort compensation funds include partial funding by the tobacco industry. Synergism's challenge to legal concepts is, however, contained by the difficulties in identifying such interactions.

4

Proof of Negligence or Breach of Statutory Duty

4.1. *Introduction*

If a disease victim can prove that the relevant hazard to health was the cause of his disease, can sufficiently fix the time of exposure (often difficult with diseases of variable latency) so that the party in control of the hazard at the relevant time can be identified, and if his claim is not barred by the statute of limitations, he must then show that the defendant is liable to compensate him for his damage. In most situations recovery in tort will depend on proof of negligence or breach of statutory duty. These form the subject-matter of this chapter, which focuses on the issues of particular importance in disease cases.

Both medical causation and breach of duty are characteristically very difficult to prove in disease cases. This is particularly so for 'test cases' where no previous case has established that the disease was, or even could be, due to the relevant drug, working conditions, environmental pollutant, etc., or was a known hazard during the relevant exposure period, and, as a consequence, such cases often require extensive and costly preparation. It is, however, a characteristic of the disease area that victims often have roughly equivalent cases against the defendant and, at least in theory, this should lead to considerable advantages. Firstly, it strengthens the precedent value of a test case and this can be useful either to future litigants or to the initial plaintiffs who may more easily extract a generous settlement out of court (see Sections 1.4 and 5.6). Secondly, it facilitates the consolidation of actions, because the evidence on the defendant's conduct and complex medical evidence would be common to all claims. This advantage of disease cases has been exploited to a considerable extent in the United States where examples of co-operation between plaintiffs range from pre-trial co-ordination in pharmaceutical litigation, to the formulation of cases as consolidated or class actions in situations involving products liability and environmental contamination liability. While class actions remain a remote possibility in England the other examples of co-ordination and consolidation are ways in which disease victims might contain the cost of litigation. (Such group activity has the added advantage of generating adverse publicity against the defendants who may thereby be more amenable to generous settlements.) A recent English example of co-ordinated disease litigation is the case of *Thompson* v. *Smith Shiprepairers Ltd.*[1] concerning industrial deafness. When this disease was prescribed under the Industrial Injuries Scheme in 1975 the pre-conditions for benefit were so narrow that a great many victims of this occupational disease did not qualify under the scheme. The result was that many of these people began to bring common law claims againt their present and former employers. As Mustill J. noted 'The response was so widespread that the legal process was in danger of being swamped by the flood of claims: more than 20,000 of them.' Since the issues in these cases were to a great extent the same, those involved in the litigation agreed that a selection of typical claims would be tried together to establish a set of criteria which could be thereafter used to settle claims out of court.

The fact that a victim may have been exposed over a considerable period of time to the disease-causing conditions creates particular difficulties both for fact-finding itself and for the judgment of these facts. In a claim of negligence the court must assess when and to what extent a risk of disease ought to have been appreciated—the 'knowledge' question—and the adequacy of the defendant's response to this level of knowledge—the 'response' question. This usually raises much greater difficulties than in cases of trauma because the risk is more often not obvious but insidious and concealed, and its appreciation depends on knowledge acquired through means other than common experience. The court may be forced to put great reliance on expert evidence as to what should have been known. Moreover, fault is not to be analysed at any one point in time but over a period, and this increases the complexity of both the knowledge and response questions as well as having implications for the assessment and apportionability of damages. Also the fact that the relevant period

[1] [1984] 2 WLR 522.

in cases of latent disease may reach back decades before trial severely undermines the quality of much of the relevant evidence. Finally, the typical disease case, such as drug damage or the continuing negligence of exposing a person to disease-producing conditions, may be more easily construed as negligence in design of a system than in its operation, and this can raise serious policy issues.

4.2 *The Duty to Avoid Damage to Another's Health*

It is conventional to open the discussion of liability in negligence with a description of the situations in which a duty to take care is owned. The breadth and flexibility of the *Donoghue* v. *Stevenson* neighbour principle, however, has generally been used by courts to impose such a duty except where policy arguments appear to require some limitation, often a requirement that the parties be in some proximate relationship.[2] In the area of physical personal injuries caused by affirmative negligent acts the general neighbourly duty has been imposed but liability for omissions to act is more restricted, principally because of the difficulty in identifying who should have acted.[3] Although for reasons discussed later the disease plaintiff rarely alleges fault in the creation of the hazard, but instead alleges omission of adequate control of the hazard, disease cases are not characterized by this conceptual reluctance to impose liability because it is usually clear—at least in cases where medical causation can be established—who, if anyone, should have controlled the hazard. This may be because that party created the hazard (e.g. a product manufacturer or polluter), knowingly handled it later in the distribution process (e.g. a blood bank, an industrial user of chemicals, a prescribing doctor, etc.), or was otherwise under a clear obligation to control the hazard posed to the plaintiff (e.g. those involved in the design, approval, construction, or occupation of premises incorporating hazardous materials; the generator or disposer of toxic waste and the occupier of the disposal site).

There are some problem areas, however. One example is infectious disease. In some situations there will still be no difficulty in pin-pointing the relevant party under a duty to take care. For

[2] *Donoghue* v. *Stevenson* [1932] AC 562. See e.g. *Anns* v. *Merton London Borough Council* [1978] AC 728, 757 (per Lord Wilberforce).

[3] Atiyah *ACL* 70, 102, 110.

example, local and hospital authorities, individual doctors and nurses can be held liable for cross-infection to another hospital patient, to hospital visitors, and to certain third parties such as the family of a prematurely released infectious patient.[4] Similarly, those who control infectious animals and organisms have a duty to prevent them escaping and infecting other animals, including human beings.[5] But infectious diseases, like environmental contamination of food-chains, can have epidemic results, and it is unclear how far courts would hold the party responsible for the initial contagion or contamination liable for the catastrophic results. Even though the courts might want to avoid the potential multiplication of litigation recognition of such a duty might entail, in practice they might be swayed by compensatory policies in those few cases where medical causation could be traced and proven.

A second area of difficulty and doubt is the extent to which an individual has a duty to 'rescue' another endangered by a health hazard. No doubt there is a duty on a person who knows he has a serious infectious disease not to cause injury to others, say, by continuing to work or to donate blood which he realizes may be infected. Where the individual is not in control of the risk but is aware of it the extent, if any, of his duty to warn those in danger depends on the general complex policy issues surrounding the duty to rescue. Traditionally, courts resist recognition of a duty to rescue but this has met with criticism, particularly where effective action requires little effort, and it seems many courts would be eager to seize on any pre-existing relationship between the parties to overcome the conventional rule. But when, and on what basis, this might be done is still unclear. In the United States, for example, a company which discovered but did not inform one of its employees that he had tuberculosis was held liable to the employee's wife who was subsequently infected.[6] Finally, one area of previous doubt which is theoretically important in the disease area—whether a child can sue for injuries inflicted before birth—has now been resolved in the affirmative by statute.[7]

[4] *Heafield* v. *Crane, The Times*, 31 July 1937; *Lindsey CC* v. *Marshall* [1937] AC 97; *Vancouver General Hospital* v. *McDaniel* (1934) 152 LT 56; *Evans* v. *Liverpool Corp.* [1906] 1 KB 160.

[5] e.g. *Weller* v. *Foot and Mouth Disease Research Instit. Ltd.* [1966] 1 QB 569, *Hales* v. *Kerr* (1908) 2 KB 601 (infection from barber's instruments).

[6] *Wojcik* v. *Aluminium Co. of America* (1959) 183 NYS 2d 351.

[7] Congenital Disabilities (Civil Liability) Act 1976, c. 28, s. 1. See Law

4.3 *Negligence in Fact*

The conclusion that a defendant has been negligent in fact—that is, fallen below the standard of care expected of him by the court—is a value-judgement on the reasonableness of his conduct in the circumstances of the case. This judgement—specific to the facts of the case and not strictly binding in later cases—involves a weighing of interrelated issues broadly separable into a consideration of what the reasonable defendant ought to have known of the risk his conduct posed (the 'knowledge' or 'foreseeability' question), and of how adequate the defendant's response to this knowledge was in view of factors such as utility and costs (the 'response' question). In a disease case where the medical causation link to an identifiable defendant can be established, the victim is typically suffering from exposure to a hazardous man-made or natural product. In some cases the disease will be due to some impurity introduced into the article during the manufacturing process. The dermatitis case of *Grant* v. *Australian Knitting Mills*[8] is a well-known example of such cases, which are generally described as 'manufacturing defect' cases. Although the complexity of modern manufacturing processes often makes detailed explication of the defendant's fault by the plaintiff virtually impossible here, courts are generally willing to infer negligence in such cases, so that the more convincing is the evidence that quality control was 'foolproof', the stronger the inference that an employee must have been negligent.[9] The practical effect is that liability for manufacturing defects in practice approaches strict liability. More typically man-made disease cases involve hazards inherent in the nature of the man-made or natural product. They are cases of so-called 'design defects'. Well-known examples are thalidomide, industrial dusts such as asbestos, consumer products like the Dalkon Shield, and environmental pollutants such as airborne lead. Here the allegation is that the creator or controller of such inherently dangerous

Commission, *Report on Injuries to Unborn Children* (Law Com. No. 60) Cmnd. 5709 (1974).

[8] [1936] AC 85.

[9] C. Miller and P. Lovell, *Product Liability* (London, Butterworths, 1977), 189; G. Schwartz, 'Understanding Products Liability' (1979 67 *Calif. L. R.* 435; J. Steiner, 'Putting Fault Back into Products Liability' (1982) 1 *Law & Phil.* 419, 442.

substances was negligent, given the actual knowledge of the defendant or the knowledge he ought reasonably to have acquired of the hazardous nature of the product design.

4.4 *The Knowledge Question: Foreseeability of What?*

For liability in negligence to arise the defendant must have responded inadequately to a risk of which he knew or ought to have known. Traditionally it is said that the risk must have been 'foreseeable', but since the knowledge question is pivotal in some disease claims it is crucial to have clear what exactly is the kind of knowledge which is involved here. There would seem to be at least four overlapping aspects of knowledge of risk:

(i) certainty . . . was the risk certain or merely suspected?
(ii) specificity . . . did the risk appear as a general risk to health or as a specific risk of, say, cancer?
(iii) probability . . . how high did the threatened percentage incidence of the disease appear to be . . . for example, one per cent of those exposed or ten per cent?
(iv) severity . . . how severe did the consequences appear to be . . . for example, was the cancer risked treatable or invariably fatal?

As an illustration take the first aspect, certainty. Say it is known that an ascertainable number, *X* per cent, of children born to mothers who took drug *Y* suffer a specific and severe deformity *Z*. Then at a particular time it may be the general consensus that these deformities were due to the drug. The risk of an *X* per cent incidence of deformity *Z* was, in this sense, regarded as certain. At an earlier date the causal link to those deformities may have only been suspected or been a matter of dispute between experts.

Even though rarely quantifiable, these factors as broad, variable, and interdependent considerations provide the sort of detail to the concept of foreseeability which allows considerable flexibility in treatment of cases. Courts can vary the individual requirements of the knowledge which must be foreseeable, including the precision with which damage is defined, in order to reach a result generous to plaintiffs. This can be illustrated by the following brief survey of successful disease cases where the risk was held to be 'foreseeable' despite the fact that one aspect of knowledge was low.

(*i*) *Low Level of Certainty* In *Wright's* case,[10] a chemical manufacturer was held negligent for failing to act on a *mere suspicion* that a chemical could cause a significant incidence of a specific and grave disease, cancer. The possible 'disastrous' consequences of the chemical outweighed the fact that there had been no cases of cancer reported in humans from either the chemical itself or related compounds.

(*ii*) *Low Specificity* In *Franklin*,[11] regulations were held to be such as to put a reasonable employer on notice that there was a risk to health from proximity to a certain dusty process, which was both clear and serious even though the nature of the lung disease was unspecified. In *Bradford*,[12] it was sufficient that the plaintiff had been exposed to a clear hazard to health from extreme cold even though the specific condition which resulted—frost-bite—was (in England) highly unusual. In doctrinal terms it was of the type and kind of injury which was reasonably foreseeable. Thus, even if a chemical is not known to be hazardous itself, but is known to belong to a hazardous family of substances, a risk may be sufficiently foreseeable.[13] In *Ransom*,[14] where there was clear knowledge of a risk to health from faulty decompressions, there was held to be no need for specificity between the risk of 'bends' and bone necrosis since the risk in both cases would have been reduced by slower decompression. Of course, specificity would be important where the health risks have to be met with different precautions.[15]

(*iii*) *Low Probable Incidence* In *Stokes*[16] the defendant's factory doctor was aware of the risk of scrotal cancer from mineral oil but considered the risk of any particular worker contracting the disease was too remote to warrant the precautionary periodic

[10] *Wright* v. *Dunlop Rubber Co. Ltd. and ICI Ltd.* (1972) 13 KIR 255.

[11] *Franklin* v. *Gramophone Co. Ltd.* [1948] 1 KB 542.

[12] *Bradford* v. *Robinson Rentals Ltd.* [1967] 1 WLR 337.

[13] *Wright's* case, *supra*; but see *Graham* v. *Co-operative Wholesale Society Ltd.* [1957] 1 WLR 511.

[14] *Ransom* v. *McAlpine (Sir Robert) & Sons Ltd.* (1971) 1 KIR 141.

[15] e.g. *Harvey* v. *New Zealand Shipping Co.* [1955] 1 Lloyds LR 251 (dust irritation hazard to eyes from cement well known but not dermatitis risk); *Tremain* v. *Pike* [1969] 1 WLR 1556 (precautions against food poisoning and bites different from those against risk of skin contact with rats' urine).

[16] *Stokes* v. *Guest, Keen & Nettlefold (Bolts & Nuts) Ltd.* [1968] 1 WLR 1776.

examinations of workers recommended in medical and government literature. This literature showed, however, that although the threatened incidence of this cancer was low, the mortality rate of victims was high. The court therefore concluded that the doctor—who had himself dealt with two cases, one fatal—had culpably underestimated the risk.

(*iv*) *Low Severity of Consequences* Finally, where there is a clear risk of disease it may be immaterial that it is a relatively minor ailment such as dermatitis, for example, where the cost of precautions is low.[17]

The preceding illustrations of the complexity of the knowledge issue in disease cases were chosen from those few disease claims which have succeeded. They show that if presented with sufficient suitable evidence courts are willing to handle the issue generously, holding that the plaintiff has succeeded on the point even though the foreseeability of that aspect of the risk was particularly low. However, useful though these cases are in showing the complexity of the notion of 'knowledge of risk' in disease cases and the lenient attitude of courts, it should be stressed that they are not representative nor numerically significant overall. It is far more typical that proof of the knowledge issue presents a formidable, often insuperable barrier in disease claims. This is because a number of factors complicate the courts' task in assessing what knowledge the defendant ought to have had. Knowledge of disease risks tends to accumulate slowly and haphazardly (see Section 3.2), so that it may be difficult to present cogent evidence on, and therefore to assess the extent of knowledge at any given point during the relevant exposure period. This is exacerbated in latent diseases because the exposure period may be decades earlier and the reconstruction of what was then appreciated about the risk may be correspondingly much more difficult. In addition, the source of a disease is often not obvious to common sense—as would be, say, the danger from an unfenced cutting machine—and this means that much will depend on what external sources of information the court judges the defendant ought to have used. This is turn requires a weighing of expert evidence which is often

[17] *Pellici* v. *Jeyes Sanitary Components Ltd.*, Guardian, 15 Jan. 1964, 15; *Voller* v. *Schweppes (Home) Ltd.* (1970) 7 KIR 228.

more available to the defendant.[18] The heavy reliance on expert evidence also increases the uncertainty of result in practice since more then depends on the forensic credibility of those giving evidence. Finally, there is an important unresolved issue here: it is not at all clear when there is a positive duty on a defendant actually to generate knowledge about the possible health risks involved in his activities. For example, it may be settled that pharmaceutical producers should carry out extensive pre-marketing testing of drugs—and post-marketing surveillance should probably also be extensive—but it is not clear what research, testing, and follow-up monitoring should be done by other creators of health hazards such as producers of chemicals like pesticides and fertilizers.[19]

Of course some disease hazards were so well known at the time of exposure that the knowledge issue is not raised at the trial. But conversely, there are many more cases where the disease risk was clearly unknown, and while in such cases a very, very few plaintiffs may be able to rely on breaches of statutory duty to establish their claims (see below), in most cases they will be without a remedy in tort.

4.5 *Adequacy of Response to Knowledge: Utility of Design Itself*

Even if a disease plaintiff has succeeded in the difficult and costly process of establishing that the defendant knew or reasonably should have known of the relevant risk, he must then show that the defendant's response was inadequate given the utility of the defendant's activity and the burden of taking precautions (including the feasibility of potential victims taking precautions). In theory, this demands a complex cost-benefit analysis, but in practice, plaintiffs rarely argue and courts rarely attempt to address such questions. Thus, in manufacturing defect cases the fact that the product did not conform to the intended design is usually taken as evidence that it was produced negligently. By adopting the article design as the standard below which no article may fall, the court simplifies the problem by requiring a foolproof manufacturing

[18] The Law Commission and Scottish Law Commission, *Liability for Defective Products* (June 1977) Cmnd. 6831, para. 36; J. Fleming, 'Drug Injury Compensation Plans' (1982) 30 *Amer. J. Comp. Law*, 297, 307; Pearson, para. 1226.

[19] A little guidance is given by *Wright* v. *Dunlop Rubber Co. Ltd. and ICI Ltd.* (1972) 13 KIR 255, and on duty of industrial user see *Graham* v. *Co-operative Wholesale Soc. Ltd.* [1957] 1 WLR 511.

process, and avoids the complex problem of whether, say, a one per cent failure rate in production is reasonable given the benefits of the defect-free product.

In the typical disease case, injuries due to design, the problem is more acute because there is usually no such external standard by which to judge the product. A plaintiff may want to argue that the product's foreseeable risks so outweighed its utility that it should not have been created at all. The question in such claims of negligent design, then, becomes how much inherent risk in a design is acceptable. In rare cases it may be possible to show that the product had no utility at all for its stated use. This may have been the case, for example, with the drug DES (diethylstilbestrol), which studies in the early 1950s indicated was ineffective to prevent miscarriage, but which was prescribed for two decades afterwards for such use.[20] Similarly, the performance level of the design may be so low that the conclusion that it was unreasonably dangerous can be based on common sense. Sometimes evidence that a product had been banned in another country or had easily available safer substitutes might enable this central question to be bypassed. In most cases, however, the court dealing with an allegation of negligent design would be faced with a broad range of value-laden interdependent factors—in a so-called 'polycentric' decision—the resolution of which involves the sort of choice between conflicting and fundamental social values which Calabresi and Bobbitt have termed a 'tragic choice'—here the essentially first order choice of how much health should society aim to protect.[21] How, for example, is a court to balance the *economic* advantages of substances such as urea formaldehyde and asbestos insulation against the risk to *health* that they pose? The problem has led some commentators to argue that—quite apart from the technical inexpertise of courts to weigh expert evidence on factual risk assessment—the adjudication process is an inadequate and inappropriate technique for resolving such open-ended value-judgements.[22] Whether or not assessment of the reasonableness of design is beyond the limits of adjudication in this general sense,

[20] R. Vagley and M. Blanton, 'Aggregation of Claims' (1981) 16 *Forum*, 636, 639.

[21] *Tragic Choices* (New York, Norton & Co., 1978), 19.

[22] J. Henderson, 'Judicial Review of Manufacturers' Conscious Design Choices: The Limits of Adjudication' (1973) 73 *Col. L. R.* 1531, 1544–7.

there is a related but more pressing reason why courts would be reluctant to address a full cost-benefit analysis head on. This is because of the much more dramatic social ramifications a finding of unreasonable design can have in certain cases. If a defendant is held liable for a manufacturing defect he can choose to absorb the cost in producing the article. If a defendant is held liable for 'defective design' on the other hand he might be forced to cease production of that design altogether. In some cases this will not have a dramatic impact because, for example, the product is not in wide use, is not an essential item, or has close substitutes. In other cases—say the asbestos, tobacco, or nuclear power industries—a finding that the product risks, even if subject to the best feasible controls, outweighed its utility could result in the collapse of an entire industry. In such cases, where value-choices inherent in a finding of negligent design could have such dramatic social impact, any judicial reluctance to make such findings can be seen as an attempt to preserve the constitutional separation of powers: that the setting of broad social priorities is more appropriate to government than to the adjudication process.

Occasionally courts rationalize a reluctance to consider the utility of design by reliance on positive government action, assuming, for example, that absence of a legislative ban on the creation of a particular health hazard leaves victims 'without . . . any legal ground of complaint . . .', 'though it may be a matter of wonder that such an operation should be permitted'.[23] The problem with this line of reasoning is that, although some processes may have attracted such a legislative ban, government agencies cannot be expected to have evaluated the utility of every hazard to health. In other words, regulatory inaction cannot be taken as tacit approval of the creation of a hazard such that a plaintiff cannot argue negligent design. Moreover, even when a government body has investigated a particular product or process and has not prohibited its creation—for example, it may grant a product licence to a pharmaceutical—it could still be argued that the judgement was not conclusive. The assessment could, for example, have been negligently made by the agency itself; may be biased because the evidence on which it is based was not independent and balanced but substantially derived from or

[23] *Nicholson* v. *Atlas Steel Foundry & Engineering Co. Ltd.* [1957] 1 WLR 613, 616 (per Vis. Simonds).

funded by the party to be regulated, or have become outdated by later acquired knowledge of risks or the development of safer alternatives. This latter point, coupled with the inadequate resources of government monitoring agencies, goes to explain why even the pharmaceutical industry itself does not regard official approval by licence as conclusive or as diminishing their responsibility.[24] The agency may itself have perfunctorily rejected a ban, explicitly emphasizing its inadequate resources to assess the broad social impact of such a decision—an attitude illustrated by the Advisory Committee on Asbestos, which then shifted back to the asbestos industry the duty to assess the narrower cost-benefit question of the acceptable uses of asbestos in the light of possible substitutes.[25] The more cogent rationale, therefore, for any judicial reluctance to consider a product design's utility is not the fiction that Parliament *has* considered the matter but that if anyone is to do so it *should* be Parliament or some public body, because it can entail weighing values of a broad social impact which courts feel neither competent nor constitutionally empowered to deal with.

This reluctance is not a hard and fast technical rule but a policy question of degree. Were a range of negligent design claims to come to court one would expect courts to be more willing to attempt a cost-benefit analysis when the social impact of a finding of negligent design would be limited. In practice, however, claims of negligent design are themselves rare.[26] While this may in part be due to lack of imagination by counsel and to a realization that such open-ended claims are both expensive to mount and liable to rejection by courts too cautious to tackle them, it is more simply explained by the fact that 'defective design' claims are usually unnecessary in the context of the individual claim. If an attack is made not on the creation of the hazard but on the inadequacy of the specific measures taken to control it, the issues in dispute become more focused and the impact of a finding of liability narrowed so that the encroachment on issues of broad social impact is minimized. In such a context courts are not reluctant to

[24] Pearson, para. 1260; S. Waddams, *Products Liability*, 2nd ed. (Toronto, Carswell, 1980), 65.

[25] Health and Safety Commission, *Asbestos: Final Report of the Advisory Committee* (HMSO, 1979) paras. 149–53, 212–13.

[26] Miller and Lovell, *Product Liability* 208 (UK); Henderson, 'Judicial Review of Manufacturers' Conscious Design Choices' 1559, 1568 (US).

impose liability. Indeed, they are able to impose an exacting standard, for example, by seeing it as doubly incumbent on a disease hazard creator to safeguard those exposed against any additional risk. Not surprisingly, then, most cases on the dangers inherent in a product or process proceed on this basis. Thus, for example, in a typical occupational cancer case the question of whether the carcinogen should have been used at all is not raised, the issue being one of adequate warnings and safeguards. The drawback to this pattern in disease case-law is that, while it can simplify the issues in some individual cases, it drastically undercuts the precedent value of any successful claim by emphasizing the particular factual context of the relevant exposure. If, in contrast, cases were to be brought where, despite all possible safeguards and warnings, the plaintiff still succeeded because the product was held to be a negligent design, this would clearly be of considerable benefit to others injured by that product. The understandable reluctance of individual plaintiffs to incur the cost and risk of a 'defective design' claim, and of courts to risk encroaching on policy areas outside their perceived competence and jurisdiction, constitute an intrinsic inhibiting factor in the development of broad principles of liability in disease case-law.

4.6 *Standard of Care in Controlling Health Hazards*

Courts have required those who control a health hazard of which they know, or ought to have known, to take certain precautions. These fall broadly into three overlapping categories: provision of safety equipment and conditions; giving of warnings and directions for handling the source of danger; and monitoring the results of exposure. Before looking at each of these it is important to have in mind the fact that, quite apart from these liability issues—the policy dilemma involved in cost-benefit evaluations of the adequacy of the defendant's conduct—there are great evidentiary problems in latent disease cases in establishing those threshold facts of what the relevant conditions surrounding exposure had been like many years, often decades, before trial.

(a) Safety Equipment and Conditions In a few cases the appropriate safety device is self-evident or spelt out in sources of information of which the defendant ought to have been aware, such as government notices on dermatitis prevention. More often

the type of precautions necessary to contain the hazard adequately are not so clear-cut. Industry practice may be a guide—since it often reflects the general level of knowledge about the risk—but it may itself be inadequate because, for example, the industry is dominated by a few concerns with poor standards.[27] Industry practice may also be irrelevant because the defendant had particular reason to doubt its adequacy or efficacy.

One way plaintiffs might be able to short-circuit the complex question of 'How much control is enough?' is to cite as evidence of negligence failure to meet advisory government standards (breach of a statutory *duty* is dealt with below, see 4.7), preferably stating a concrete precaution such as maximum concentrations of airborne substances in workplaces. The problem is that relatively very few disease hazards can, and have been, publicly assessed in this way, and many of those numerical standards which do exist are based on inadequate data. Virtually all apply to the industrial context—notably the 'threshold limit values'—where the variety of working conditions undercuts the evidentiary value of failure to comply. Defendants could argue that the advisory levels were too low both because of limitations in the accuracy of risk assessment (for example, dose–response relationships, etc.—see 3.4, 6.9), and because the cost-benefit valuation of those assessed risks was flawed. An example of the latter may be the cotton industry, where many mills at present fail to comply with advisory maximum dust concentrations because, it is argued, compliance costs are beyond the resources of this contracting industry.[28] The inherent polycentric problem of assessing benefits and costs of certain numerical standards is well illustrated by an American study where seventy-two estimates of the two-fibre standard for asbestos produced benefit/cost ratios ranging from 0.07 to 27.70—a 400-fold range.[29] In assessing a standard the regulatory agency may itself have explicitly acknowledged that setting a fixed acceptable level of risk (say, two per cent excess mortality) would be arbitrary as 'society has not yet developed any method or basis for defining

[27] e.g. the asbestos industry, where major producers have been attacked for substandard hygiene practices: *Smith* v. *Central Asbestos Co. Ltd.* [1971] 3 All ER 204, 208 (Lord Denning). See also *Thompson* v. *Smith Shiprepairers Ltd.* [1984] 1 All ER 881 on industry-wide apathy to noise.

[28] P. Neild, *Byssinosis: The Lancashire Disease* (London, Chart. Insur. Instit., 1982), 5, 42, 45.

[29] J. Mendeloff, *Regulating Safety* (Cambridge, MIT Press, 1979), 63.

acceptibility in this sense'.[30] In any case, since there is no mandatory duty adequately to monitor exposure levels, in many cases there will be no record or only an inadequate record of failure to comply with advisory standards and, especially in long latent diseases, current hygiene levels are usually of little evidentiary weight.

In the absence of external standards the court is reduced again to assessing the cost-benefit question of adequacy of precautions, of how much risk is socially acceptable, and of what level of disease incidence from a particular source is reasonable. A favoured technique by which this can be avoided is to narrow the question to the issue of adequate warnings.

(*b*) *Health Warnings and Directions for Use* Typically, disease risks are not obvious to the observer. Toxic dust, chemicals, and invisible radiation may not give immediate, palpable injury, such as burns, but act imperceptibly and insidiously on the body. The imposition of a duty to give detailed, vigorous warnings and directions on how to handle the source of such risks is, therefore, a typical focus of disease case-law. The cases are dominated by predictable issues such as the nature and intensity of the necessary warning given the level and concealment of danger, and whether or not it is sufficient to warn an intermediary such as a prescribing doctor.

Where there is long term exposure to a health hazard—the typical situation in the occupational context—it is sometimes argued that as those exposed gain experience the necessity for stringent warnings and supervision becomes less. But, while this view may be appropriate to the sort of cases of accidential injury where it was developed—that is, where the risk was obvious and the results immediate—it is unsuited to the risks associated with most diseases. Here the risks are insidious, a factor which even in accident cases is used to impose stringent warning requirements.[31] Moreover, with latent diseases the passage of time without apparent ill effects from exposure misleadingly lowers consciousness of risk and may in fact call for continued, if not increased, efforts to warn of the risk.

[30] *Asbestos: Final Report of the Advisory Committee*, para. 162. See also para. 213.

[31] See e.g. the comments of Lord Keith in *Qualcast (Wolverhampton) Ltd.* v. *Haynes* [1959] AC 743, 755; *Bux* v. *Slough Metals Ltd.* [1974] 1 All ER 262, 273 (per Stephenson LJ).

(*c*) *Monitoring Results of Exposure* Since disease hazards are typically concealed, insidious, and latent in manifesting damage, an area of particular interest is how far a party who has created or otherwise had control of the hazard is required by the common law to monitor and investigate the effects it has on those exposed. Despite its importance, this is an area of considerable doubt because of the paucity of relevant case-law in England. In particular, it is unclear how active this collection of information need be (is continuing research necessary?); how much reliance can be placed on public monitoring bodies (for example, adverse drug reaction committees); whether warnings of subsequently discovered risks should be made to all past as well as future users; when, if ever, production must be halted; and how extensive, if at all, is the duty to retrieve dangerous products already in circulation in this context.[32]

The continuing and close relationship of employer and employee adds another dimension to this question of monitoring the results of exposure to a health hazard. In some cases an occupational disease victim may argue that his employer, even though not liable for the creation of the hazard, was under a duty to monitor his employees' health in such a way as to detect early signs of the disease, thereby limiting its progress. However, the British Medical Association sees ethical problems in pre-symptomatic screening, specifically where no practical assistance is scientifically or economically feasible at this earlier stage,[33] and it raises serious issues in the area of discriminatory employment practices (see below).

4.7 *Breach of Statutory Duty; Compliance with Advisory/ Mandatory Standards*

One way disease plaintiffs may be able to avoid the open-ended cost-benefit problem of proving that the defendant's design or control of the health hazard was inadequate is by showing that the defendant was in breach of a relevant statutory duty. In theory, the most useful form such a statutory crystallization of the

[32] There is more case-law on these points in the US: see: Miller and Lovell, *Product Liability*, 233, 243, 248 ff.; Waddams, *Products Liability*, 48–9. See also footnote 19 above.

[33] British Medical Association, *The Handbook of Medical Ethics* (London, BMA, 1981), 31.

standard of care can take is a mandatory numerical limit on exposure to, say, coal dust, radiation, lead, noise, weights, pressure, etc. But such standards are very few, and in practice, lack of adequate monitoring records would make proof of breach impossible in virtually all cases. One notable exception is *Ransom* v. *McAlpine (Sir Robert) & Sons Ltd.*[34] but, significantly, the relevant regulations here required monitoring and the keeping of records.

More common are non-numerical standards or duties such as those relating to the unlicensed disposal of poisonous waste, the state of buildings, and in particular to industrial health (for example, labelling of toxic industrial chemicals; provision of protective clothing and equipment; monitoring exposure levels and the health of those exposed; and warnings to those exposed). The most well-used of these are ss. 4 and 63(1) of the Factories Act 1961, which deal with the control of fumes, dust, and other impurities in factory air. The second limb of the duty under s. 63(1) illustrates an anomalous advantage, from the disease plaintiff's viewpoint, of relying on certain breaches of statutory duty: here, the occupier may be in breach of duty without knowing of the injurious nature of the relevant substance. This has been exploited to establish liability for exposures occurring before a reasonably well informed party in the defendant's position would have known of the disease hazard, and although judges have complained of this 'odd and artificial' use of a duty aimed at visible dust to fix liability for damage caused by invisible dust (for this is the dust which passes through the respiratory tract and causes disease), it may be argued that the statutory duty was of a 'pace-setting' nature in that it assumes, with apparent good sense, that any dust may be injurious to health if there is a substantial quantity of it.[35] On the other hand, even with this sort of provision, the duty to take 'all practicable measures' will be assessed in terms of the reasonably foreseeable danger and, therefore, the provision of a mask which would have coped with the visible dust, but not the harmful invisible dust, would have been sufficient to discharge the duty.

[34] (1971) 11 KIR 141.

[35] *Richards* v. *Highway Ironfounders (West Bromwich) Ltd.* [1957] 2 All ER 162, 164. See also *Cartledge* v. *Jopling (E.) & Sons Ltd.* [1961] 2 Lloyds Rep. 61, 63; *Brooks* v. *J. & P. Coates Ltd.* [1984] 1 All ER 702; *Gregson* v. *Hick Hargreaves & Co. Ltd.* [1955] 3 All ER 507.

The result is that extremely few victims can exploit this particular anomaly. So overall, although there have been generous interpretations of statutory duties in the occupational disease context, there are relatively very few statutory duties applicable in the disease area, and where they exist there are evidentiary difficulties in rebutting a defence claim of compliance.

If the evidence shows or is consistent with a defence claim of compliance with advisory precautions or with mandatory statutory duties, this does not prevent the conduct being held to be unreasonable at common law. In any particular fact situation the common law duty may be broader or narrower than that advised or laid down. In some cases it will be broader because the standard of care advised or mandated has become—particularly in cases of latent diseases of gradual contraction—out of date since being written because information on the risk has increased. Similarly, for reasons discussed earlier, the judgement of regulators both as to the technical risk assessment and to the cost-benefit policy evaluation may be flawed because, for example, of the latter's negligence; or because the medical or economic evidence on which it is based is inadequate or not independent of those to be regulated. Furthermore, the defendant may himself have special knowledge which raises his duty at common law above simple compliance with statutory duty: for example, knowledge that a safety device, supplied under a simple statutory duty to supply it, is not being used. In a very few cases a statutory duty may be held to represent a pace-setting standard fixed at a level above that required at common law because, for example, of the greater knowledge and experience of government advisers. But the modern trend of interpretation seems to be a greater emphasis on the fact that by their nature—the requirements of precision and workability—statutory duties may overlap, produce inconsistencies and gaps, and 'ought not to be expected to cover every possible kind of danger' in every particular situation.[36] Compliance with advisory or mandatory standards, then, may be evidence of fulfilment of the common law standard of care, but it is not conclusive. Of course there is an advantage to plaintiffs in this approach, particularly because of the often intractable evidentiary problems in rebutting a defence assertion of compliance when the

[36] *Gill* v. *Donald Humberstone & Co. Ltd.* [1963] 3 All ER 180, 183 (per Lord Reid). Compare *Budden* v. *B.P. Oil Ltd. and Another* [1980] JPL 586.

circumstances of exposure were long past and not contemporaneously recorded. But the plaintiff must still establish what was the common law standard of care required of the defendant and, as we have seen, this can be especially problematic where hazards to health and latent insidious diseases are involved.

4.8 *Summary*

To sum up so far: even in cases where medical causation is establishable, proof of breach of duty is particularly difficult in disease cases for a number of reasons. First, apart from certain cases of industrial dust diseases, a claim of breach of statutory duty will not be available to simplify the breach of duty issue for plaintiffs. Secondly, predictably important issues in negligence claims such as the level of knowledge and adequacy of warnings given by the defendant are typically far more complex than in accident cases. This is because disease risks are insidious, because the disease is or appears to be contracted over time—so that the defendant's knowledge and conduct must be assessed over a period—and because with latent diseases this relevant exposure period is often a considerable time in the past. The evaluation of these predictable issues is, therefore, typically extremely difficult. Often complex expert evidence is required which makes the claim particularly costly, the outcome less clear and, therefore, litigation more risky. Thirdly, there are special issues raised by disease cases such as the extent of the duty to monitor exposure which, because of the paucity of case-law, remain underdeveloped and consequently difficult to predict. Fourthly, in disease cases the injury tends to be due to the product as designed yet plaintiffs virtually never allege 'defective design'. Rather the issue is narrowed to the adequacy of the control of the health hazard in the particular circumstances of the case. While this assists the individual litigant by containing costs and avoiding possible judicial reluctance in this policy area, it inhibits the growth of case-law having a strong precedent value. Because in disease cases predictable issues are particularly complex and special issues are left underdeveloped, issues of liability are often surrounded by doubt. This unpredictability coupled with the characteristic cost of such complex claims inhibits litigation which, of course, in turn perpetuates the paucity of disease case-law.

4.9 *Remoteness, Thin Skulls and Susceptibility*

Another particular feature of disease cases which disadvantages plaintiffs is that there is little room for the thin-skull rule to operate. The basis of this rule of remoteness, unconventional and anomalous in that it allows recovery for unforeseeable losses, is the simple policy choice that as between an 'innocent' victim and 'guilty' tortfeasor the latter should bear losses associated with the victim's pre-existing susceptibility. The way the rule works in accident cases is that if an overt trauma was foreseeable then the fact that the victim's pre-existing susceptibility or condition renders consequent physiological results much graver than could be foreseeable is held to be irrelevant, and the defendant is liable for those consequences. Thus in the leading case of *Smith* v. *Leech Brain*[37] a foreseeable accidental burn triggered an unforeseeable cancer. The defendant was held liable for the cancer. The burn provides a hook on which to hang the unconventional analysis, but it is a technique for expanding liability for personal injuries which cannot be exploited in disease cases because in those cases there is no trauma on which to seize as a focus for the knowledge or foreseeability judgment. In order to answer the negligence in fact question the court is forced to ask the conventional foreseeability question:[38] was the contraction of this disease itself, that is, of the injury which is complained of and for which damages are sought, foreseeable [or a disease which is caused by the same source and could be controlled by the same precautions[39]]. Included in this question is the foreseeable incidence of the risk eventuating (see issue (iii) above) which in such a disease case is the apparent incidence of susceptible people. Consequently, in the typical disease case the negligence in fact issue itself must encompass issues such as the abnormality of the plaintiff both in terms of the incidence of susceptible people to his disease and the likely gravity of consequences. Once negligence in fact is found there is no issue left for the thin-skull rule to work on, and it therefore does not assume importance in disease case-law (except to accommodate

[37] *Smith* v. *Leech Brain & Co. Ltd.* (1962) 2QB 405.

[38] See *Overseas Tankship (UK)* v. *Mort Dock & Engineering Co.* [1961] AC 388.

[39] If there are two risks, and the precautions against one are effective to some extent against the other, the court may take this into account by requiring less specificity in defining the risk whose foreseeability etc. is to be analysed. See footnote 13 and accompanying text.

the fact that the course of a disease such as pneumoconiosis is more disabling in some victims than in others).

There is an important practical consequence of these rules. In a *Smith*-type case the presence of a foreseeable trauma, the burn, enables the court to hold the defendants liable under the thin-skull rule for unforeseen cancerous consequences, whereas in non-traumatic cancer cases liability will be far more restrictive depending on the complex issues of foreseeability of the cancer. Borderline cases highlight the dramatic shift towards broader liability which can be affected by the simple fortuitousness of there being a foreseeable, albeit not serious, trauma on which to hang the rule. Take, for example, the case where foreseeable abrasions have given rise to severe dermatitis. The court might treat it as a *Smith*-type case, holding the defendant liable for the dermatitis even if unforeseeable, or it could require that the disease itself had been foreseeable. In fact, in the few such borderline cases reaching the courts, the latter approach is taken: requiring the disease to be foreseeable where the initial precipitating physical injury is relatively minor such as blistering or abrasions.[40] While this is, of course, a significant problem for the individual plaintiff, the principal point to stress here is the much broader one that the virtual unavailability of the thin-skull rule puts disease victims as a class at a considerable disadvantage compared to accident victims.

If susceptibility is not discussed in terms of the thin-skull rule in disease cases then where does it enter the analysis? Unlike traumatic injuries, where it is almost invariably the case that the victim's physiological state plays no role in the occurrence of the accident (as opposed to the degree of harm thereby suffered), in disease cases it is a truism that the victim would not have contracted the disease but for his susceptibility. Susceptibilities vary; some individuals will be affected by a particular hazard and to varying degrees, others will not be affected at all. Consequently, in a disease case, susceptibility may be a potential cause in fact of the injury if regarded as sufficiently 'abnormal'. It is, however, rarely handled as such and in the case-law it is far more common for it to be subsumed under the broad negligence in fact issues, going both to (*a*) foreseeability of risk, and (*b*) adequacy of the defendant's

[40] *Pellici* v. *Jeyes Sanitary Components Ltd.*, *Guardian*, 15 Jan. 1964, 15; *Riddick* v. *Weir Housing Corpn.* 1971 SLT 24.Compare the process/trauma distinction, Chapter 3.

response. (*a*) If very few people are susceptible to a disease under the relevant conditions, there will have been few cases to put the defendant on guard to the risk (although where susceptibility is low and there are no previous cases, other factors may put the defendant on notice). (*b*) In taking precautions to control a disease hazard the defendant must take account of the common infirmities of individuals, even though he is ignorant that a particular person has an infirmity. The matter is one of degree and judgement: where a toxin is known to cause disease in a large *minority* of exposed persons, the fact that any particular victim's susceptibility may strictly be described as 'abnormal' will be irrelevant and precautions will be called for. In contrast, where the toxin is known to cause disease in only a small number of people exposed, the precautions required may be relatively undemanding such as warnings on product labels.

An interesting point arising from a case on susceptibility to industrial dermatitis was the judge's suggestion that, had pre-employment screening for abnormal susceptibility been possible by a simple test, failure to carry out such tests may have amounted to negligence.[41] Were this suggestion to be developed in later cases it would prove a double-edged sword for plaintiffs—admittedly advantaging them when screening had not occurred, but raising difficulties where it had been refused by the victim or undergone giving a positive result for susceptibility. This raises the issue of the reasonableness of the victim's own conduct in subsequently accepting employment, although there are also concomitant implications for the standard of care expected of an employer who is acquainted with the particular susceptibility of an employee (see below). The issue also raises general problems regarding discrimination and the employment rights of workers known or suspected to be abnormally susceptible. It is not unexpected, therefore, to find that the court have been reluctant to set up a common law duty to screen for susceptibility,[42] and there is a similar reluctance to legislate for pre-employment screening. Similar policy issues also complicate the question of whether

[41] *Darvill* v. *C. & J. Hampton Ltd.* (1972) 13 KIR 275, 284 (Rees J.).

[42] See e.g. *Joseph* v. *Ministry of Defence, The Times*, 4 Mar. 1980; *Parkes* v. *Smethwick Corp.* (1957) 121 JP 415. On US position see M. Rothstein, 'Employee Selection Based on Susceptibility to Occupational Illness' (1983) 81 *Mich. L. R.* 1379.

regular monitoring of those exposed to a health hazard is reasonably to be expected (see 4.6 (*c*)) and the implications of a person refusing such surveillance.[43]

4.10 *The Victim's Conduct*

Contributory Negligence Allegations of contributory negligence are rarely made in reported disease cases.[44] Often this is because the victim was simply not in a position to avoid the risk, for example, from food additives or environmental pollution—in economic parlance, 'unilateral' situations (see 6.4). There are, however, some disease hazards which can be reduced or avoided by the conduct of the potential victim, for example: by availing himself of medical screening/monitoring services; by using safety equipment such as protective clothing and masks; by obtaining and heeding reasonable medical advice such as to stop smoking or working in a dusty environment; or by following the directions for use of a product such as drug dosages or patch testing for cosmetics before use. Similarly, a person who knows he is susceptible to a hazard may be contributorily negligent for failing to protect himself adequately. Even so there appears to be a marked reluctance by the courts to find contributory negligence in disease cases, a feature probably best explained by the insidious nature of the typical disease hazard and the fact that the defendant is usually in a better position to appreciate and take measures to control the risk. Thus, for example, in *Smith* v. *Central Asbestos Co.*[45] the allegation of contributory negligence was rejected on the basis that the defendants had not sufficiently brought home to the employee the importance of wearing respirators.

Where attacks on the victim's conduct are made in disease cases they characteristically take the form of causal arguments, not contributory negligence. There are two types of argument here: first, the simple case where the defendant argues that the victim's failure, for example, to use a provided safety device such as a face mask or follow product directions, was so unforeseeable that it rather than the defendant's conduct was the real cause of his disease; secondly, where the defendant argues that even if he had

[43] e.g. *Asbestos: Final Report of the Advisory Committee*, 238.

[44] One of the few examples is *Clifford* v. *Challen (Charles H.) & Son Ltd.* [1951] 1 KB 495 (50% reduction for failure to use barrier cream).

[45] [1971] 3 All ER 204.

provided an *omitted* safety device, the victim would not have used it. Framing the attack in these ways has important advantages for the defendant. First, the onus of proof is on the plaintiff to rebut such causal arguments and, especially in the latter case where he must show affirmatively that he would have used the device had it been provided, this may be extremely difficult (see Chapter 3). Secondly, if the plaintiff fails on this point then his entire claim fails outright and he recovers no damages at all. The harshness of these arguments is sometimes ameliorated by the courts requiring a higher standard of care, for example, by requiring the defendant not only to have supplied the safety device, but to have taken into account human inadvertence, and to have warned the victim so vigorously that on the evidence it was likely the victim would then have complied (see Section 3.5). However, this approach is also unsatisfactory as it is simply the other side of the 'all-or-nothing' causal coin. The defence argument in both cases reintroduces the illogic and injustice inherent in the common law's refusal to recognize that the injury may be caused by both the defendant's and victim's fault. It is tantamount to an assertion that the victim had the 'last opportunity' to avoid the risk and his failure to do so rendered him the sole author of his injuries. It was to wipe out such crude 'all-or-nothing' causal arguments that apportionment was introduced. That these harsh rules can still surface so damagingly for plaintiffs in disease cases is yet another result of their preference for arguing their cases on the narrower basis of inadequate control of the hazard (that is, negligent omission). It is another risk incurred as the price to be paid in order to avoid the costliness of an allegation on the broad basis of negligent design (that is, negligent creation of the hazard).

Finally, two particular issues concerning the plaintiff's conduct which are raised by disease cases should be mentioned. Firstly, failure to inform a relevant party of a known physical susceptibility may amount to contributory negligence.[46] But here it is the plaintiff who can take advantage of an unattractive 'all-or-nothing' causal rule—requiring the defendant to prove that had the plaintiff disclosed his susceptibility the defendant would have acted to reduce or eliminate the risk. Secondly, the selectivity of most diseases raises the question of whether susceptibility which is self-

[46] e.g. *Cork* v. *Kirby Maclean Ltd.* [1952] WN 399 (epilepsy, 50% reduction).

induced—for example, by excessive smoking or drinking alcohol—is a form of contributory negligence. Presumably this might depend at least in part on how 'foreseeable' such behaviour should have appeared to the defendant, but is it also relevant that such self-induced weaknesses are themselves the result of using dangerously *addictive* products the dangers of which are insidious and arguably insufficiently exposed by current packet warnings?[47] Because of the absence of English case-law and rarity of American case-law on the point, largely attributable to problems of medical causation and identification of the relevant tortfeasor, there is no indication how courts would react to this argument, and it might seem unlikely to succeed given the difficulty in distinguishing such conduct from less dramatic forms of self-neglect such as obesity and failure to exercise. It is interesting to note here, however, that the possibility that a pregnant woman's smoking, alcohol or drug intake might constitute negligence towards her child appeared sufficiently real to the Bar Council and Law Commission to lead both to recommend the statutory exclusion of claims by a child against its mother for negligence in the antenatal regime.[48] Despite the enactment of this proposal the issue is not dead in this context either: it was inadvertently reintroduced by enabling non-parent defendants (for example, drug companies) to raise the mother's contributory negligence in a claim by the child for pre-natal injury.[49] (Indeed, the issue is more likely to arise in this context in those few cases of pre-natal injury where causation can be established.) In general, then, the prospective disease plaintiff is faced with a major uncertainty: the danger that in future cases self-induced susceptibility may be regarded as a form of contributory negligence and result in a substantial reduction in damages.

Voluntary Assumption of Risk: The Volenti Defence When the plaintiff is held to have voluntarily assumed a risk, the defendant's

[47] See Chapter 5; Miller and Lovell, *Product Liability*, 188, 211, 216–17; Waddams, *Products Liability* 45; O. Tilevitz, 'Judicial Attitudes Towards Legal and Scientific Proof of Cancer Causation' (1977) 3 *Col. J. Environ. L.* 344, 370–3; D. Garner, 'Cigarette Dependency and Civil Liability: A Modest Proposal' (1980) 53 *South. Calif. L. R.* 1423.

[48] See Law Commission, *Report on Injuries to Unborn Children*, para. 55; Congenital Disabilities (Civil Liability) Act 1976, c. 28, s. 1(1).

[49] Congenital Disabilities (Civil Liability) Act 1976, s. 1(7), criticized by Pearson, paras. 1473–7.

conduct is irrelevant. The latter's carelessness goes undeterred and the victim uncompensated. It is, therefore, not surprising that courts are hostile to the notion of express or implied acceptance of risk and have imposed stringent requirements on the *volenti* defence, specifically that the consent be fully informed, 'free', and relate not just to the physical risk but also the legal risk (that is, of being unable to sue the defendant in negligence). There is no example in the case-law of the *volenti* doctrine being explicitly applied to disease victims. In some cases the facts would have admitted the defence. It could, for example, have applied in *Withers* v. *Perry Chain Co. Ltd.*[50] where an employee, fully aware of her susceptibility, persisted in exposing herself to the hazard, but the court preferred to reject the damages claim on the basis of no breach of duty. Since virtually all disease case-law relates to the workplace, the general absence of the application of the defence is not surprising because of the difficulty in showing a fully informed and particularly a 'free' decision to accept risk under these circumstances. For example, the mere fact that nurses know of the inherent risk in working with infectious patients does not mean their decision to do so is 'free' and that they are precluded from complaining at law if they contract the disease. Indeed, the failure of courts to hold that the creation of disease hazards may be negligent *per se* makes it difficult to see when, if ever, a disease victim's conduct in exposing himself to it could be regarded as unreasonable or as a 'free' acceptance of the risk. The focus must be on the adequacy of control of the hazard, not on assumption of risk. More generally, the insidious and concealed nature of many disease risks suggests that the *volenti* defence would be extremely hard to establish in many non-employment situations, such as diseases from environmental pollution or dangerous products. Freely consorting with a person known to be suffering from an infectious disease may be one of a very few cases where the defence would succeed. Lung cancer from cigarette smoking may also be another case where there is sufficient appreciation of risk to allow the defence—although, again, difficulties in establishing full appreciation of risk and that the assumption of risk associated with an *addictive* drug is 'free' may be considerable (see above). Finally, while courts are reluctant to allow the defence to

[50] [1961] 1 WLR 1314.

exculpate the tortfeasor fully, 'acceptance' of a risk by the victim may be an important factor in lowering the standard otherwise required of the defendant where certain risks are assumed, for example, by volunteers for medical research, entrants to hospital premises, or where a terminally ill cancer patient knowingly accepts treatment with a highly toxic drug. That this does not absolve the defendant from a duty of care is clear—the doctor in the latter case, for example, must still take care in setting dosages and monitoring response to minimize the risk. Because the risk accepted is limited, so, too, is the reduction in the standard of care.

4.11 *Conclusion*

If the particular issues disease cases raise in the proof of fault (and, in a very few cases, breach of statutory duty) are examined, some characteristics of personal injuries claims are found to be less important, such as contributory negligence and the thin-skull rule, while others, such as susceptibility, take on a central importance. Disease cases highlight a number of interesting policy questions. One example is the trade-off between a neighbourly duty to monitor health with the employment rights of those to be monitored. Another is the problem intrinsic to the dangerous design cases of which most disease cases are examples, namely the practical and constitutional limits of the adjudication process. There are also important issues which disease cases could in theory be presenting to courts but which have not been significantly explored because of the problem of medical causation in the relevant cases. Examples of this include the intriguing liability issues suggested by the 100,000 premature deaths in Britain per year due to tobacco (see 1.2), and the important remoteness questions raised by environmental contamination (see 4.2).

Even when causation is established, the operation of liability rules can often disadvantage the plaintiff. For example, certain doctrines operate or are used to produce more restrictive liability than in typical trauma cases. The unavailability of the thin-skull rule and the use of all-or-nothing causal arguments against the victim are two examples. More generally the disease victim will be disadvantaged by the difficulty in predicting how liability principles will operate in this rarely litigated area. Predictable issues such as the level of knowledge of the risk and the adequacy of the defendant's warnings become particularly complex when the insidious disease was contracted gradually over time and many years before trial. Their presentation is costly and, as they often rely on stale or conflicting expert evidence, their resolution is often difficult to predict. Certain policy issues raised by the particular nature of disease, such as the duty to monitor exposure, are underdeveloped because of the paucity of case-law. Other conceptual issues, such as those involved in a claim of 'defective design', remain largely unexplored. In an individual case there is pressure to avoid such claims by using a more limited allegation of inadequate control and, while this narrowing of the liability issues to the facts of the case may assist the individual victim, it restricts the precedent value of the case. These factors combine to make the parameters of liability in disease cases particularly uncertain in crucial areas. Litigation is inhibited, therefore, not simply because of the cost of establishing the often intrinsically difficult factual issues in disease cases, but because predicting how liability rules will operate in an individual case is difficult and complex. This inhibition of litigation, of course, then tends to perpetuate the problems of an inadequate body of case-law.

Where fault cannot be proven the victims go uncompensated. In particular, this is always the case for victims of risks which were not known nor reasonably discoverable—perhaps thalidomide was one such case—and victims of risks unavoidable by reasonable means which were known but acceptable in view of the utility of the activity—arguably whooping cough vaccine is an example here. Concern that these innocent victims of socially sanctioned activity go uncompensated by tort law has been a major factor in recent calls for stricter tort liability and, in some cases, for the replacement of tort by other compensatory mechanisms. These issues are discussed in the following chapter.

5

Compensation Without Proof of Fault

5.1 *Introduction*

Chapter 4 discussed the particular difficulties in proving fault in a disease claim. This chapter begins by looking for existing areas of liability where this might not be necessary and evaluating how useful they might be to disease victims in obtaining compensation. The same question is then asked of proposals to reform the common law by making liability stricter. There are intrinsic problems in relying on tort remedies and the litigation/settlement process to provide adequate compensation even for those who might, in theory, be able to establish liability, and the particular difficulties in this area for the disease victim are discussed. One outcome of the substantive and practical barriers to success in tort generally has been a proliferation of limited non-tort schemes for the compensation of specific classes of disease victim. Although these provide better compensation for certain victims there are still both practical demarcation problems and policy objections to these *ad hoc* schemes.

5.2 *Existing Areas of Strict Liability*

Relevant existing areas of strict liability are mainly concentrated around the relationship of neighbouring occupiers of land and so what scope there is here for recovery by disease victims lies principally in the area of pollution—the escape from a piece of land, by air, water, or other means, of some disease causing factor such as radiation, dust, fumes, infectious animals, germs, etc. Unfortunately for these victims, the central cause of action here—nuisance—is surrounded by doubts as to whether it gives rise to liability for personal injuries: as the action in private nuisance arises from the unreasonable invasion of the plaintiff's interest in the beneficial use of land, it is thought to exclude recovery for personal injuries.[1] If this view is correct, the

[1] J. Holyoak and D. Allen, *Civil Liability for Defective Premises* (London, Butterworths, 1982), 56; J. Fleming, *The Law of Torts*, 6th ed. (Sydney, Law Book

neighbouring occupier of, say, an asbestos factory may be able to sue in private nuisance for dust settling on his property on the basis that the resultant health risk has reduced the usefulness of his land. He would, however, be unable to recover in private nuisance for an asbestos-related disease which he actually contracts, being relegated to the remedies available to victims without occupier status. It has similarly been doubted whether the related rule in *Ryland* v. *Fletcher* allows recovery for personal injuries. In *Read* v. *Lyons & Co. Ltd*,[2] Lord Macmillan appeared to reject such a possibility outright, but a number of cases and later commentators suggest that this *obiter dictum* should be read as simply referring to the unavailability of this cause of action to a person without the appropriate interest in property. Following this latter approach it has been suggested that the rule in *Rylands* v. *Fletcher* may provide a useful tool for those (with a sufficient proprietary interest) who suffer disease from the release of materials such as radiation, germs, or toxic waste from neighbouring property where proof of negligence is difficult or impossible.[3] Buckley has, however, strongly argued that a literal reading of Lord Macmillan's dictum is preferable, not least because it avoids the serious doctrinal anomaly of allowing the 'short cut' of strict liability to a very small minority of personal injury victims while leaving the vast majority—including, after all, the occupier's spouse, children, visitors, etc., and those injured on the original premises themselves—to prove negligence.[4]

Relatively more scope for victims of environmental health hazards is presented by the 'anomalous' area of public nuisance where interference with the beneficial use of land is not a necessary element and where consequently recovery by non-occupiers of land for personal injuries is well accepted. The gist of

Co., 1983), 380; R. Buckley, *The Law of Nuisance* (London, Butterworths, 1981), 76.

[2] [1947] AC 156, 173.

[3] J. Bodie, 'The Irradiated Plaintiff: Tort Recovery Outside Price-Anderson' (1975–6) 6 *Environ. Law*, 859, 884; Y. Cripps, 'Avenues of Compensation for Genetic Engineering Accidents' (1980–1) 9 *NZULR* 150, 155; W. Ginsberg and L. Weiss, 'Common Law Liability for Toxic Torts: A Phantom Remedy' (1981) 9 *Hofstra LR* 859, 913 ff.

[4] Buckley, *The Law of Nuisance* 48–9; Holyoak and Allen, *Civil Liability for Defective Premises* 49. See also C. Miller and P. Lovell, *Product Liability* (London, Butterworths, 1977), 253; Law Commission, *Civil Liability for Dangerous Things and Activities*, Law. Com. No. 32, 1970, para. 13.

such claims is that the defendant's use of land encroached on the safety or convenience of the public in general and has caused special damage to the plaintiff in particular. The action, therefore, seems particularly well suited to pollution disease cases—at least where the pollution caused palpable inconvenience to a sufficiently large group of people—where special damage would be constituted by the actual contraction of disease by a minority of those exposed. But there are problems here too. It is not at all clear how courts would handle the susceptibility of the plaintiff in such a context. Moreover, the case-law is divided on whether some type of fault must be proven. Given the anomalous nature of such public nuisance actions for personal injuries which do not arise from a proprietary interest, it is likely that courts would take the line suggested by Lord Reid in the *Wagon Mound II* case—that some type of fault is nearly always necessary.[5] In the result, the scope of liability approximates to that in negligence with perhaps the sole advantage of suing in public nuisance being that the onus seems to be on the defendant to prove lack of fault. While this is not an insignificant advantage, there are still considerable practical reasons why the defendants in disease claims are in a good position to discharge such an onus (see Chapter 4). In addition, statutory authority may in a few cases provide the defendant with a complete defence to the claim of nuisance (but not negligence) if it was an inevitable result of an activity authorized by Parliament.[6]

There is one area where both difficulties in the proof of fault and the anomalous requirement of a proprietary interest have been avoided in constituting a liability for environmental personal injuries. This is in the Nuclear Installations Act 1965 (c. 57) where Parliament introduced strict liability for personal injuries and property damage caused by a nuclear 'occurrence'. There is no restriction on who can sue—so that, for example, employees within the installation are covered—and liability is channelled to the licensee of the relevant plant thereby concentrating and reducing the overall cost of the compulsory insurance cover. Comparable disasters can of course occur in non-radiation areas so, for example, the disastrous potential of the escape of virulent

[5] *Overseas Tankship (U.K.)* v. *Miller Steamship Co. Pty.* [1967] 1 AC 617 ('Wagon Mound II') 639 (per Lord Reid). See also Buckley, *The Law of Nuisance*, 65–8; Holyoak and Allen, *Civil Liability for Defective Premises*, 59.

[6] *Allen* v. *Gulf Oil Refinery Ltd.* [1981] 1 All ER 353.

forms of genetically engineered organisms has led to calls in other countries for a form of strict liability to be imposed either on the originating laboratory or the government which has supported or at least authorized such controversial research. Compensation for some biological hazards may fortuitously be facilitated by the strict liability imposed by the Animals Act 1971 (c. 22)—for example, because infectious animals are in a 'dangerous condition' or because the relevant virus or bacteria belonged to a 'dangerous species'. But in general there is as yet no special device by which victims of an environmental disaster could recover compensation —even though it is in these cases where private insurance pools are most likely to run out.

In summary, there are very few areas of strict liability available to the disease plaintiff at common law and in only a few isolated areas—nuclear installations and the keeping of animals—is there strict liability imposed by statute. Moreover, strict liability solves only one of the problems facing the disease plaintiff—proof of fault—leaving issues such as proof of medical causation which are especially problematic in these environmental cases. Finally, and this is a point which will recur in the following discussion, areas of strict liability, like all systems by which compensation is facilitated for a limited group of victims be it under the common law or otherwise, produce anomalous results between cases which have provoked widespread criticism and are difficult to justify (see 5.9).

5.3 *The Risk Definition Problems in Strict Liability and No-Fault Reforms*

One way to ease the burden of the disease victim under tort law would be to widen the areas of liability where proof of fault is unnecessary, that is, areas of strict liability. In terms of the compensation goal the advantage of such reform is that compensation is received by a wider class of victim on a basis which is theoretically simpler and less costly to determine than fault. These advantages could, alternatively, be combined with those of a non-tort system of compensation from a fund, giving the 'no-fault' basis of liability of such schemes as workmen's compensation (private insurance funds) and later the industrial injuries scheme (state fund). Given the often widespread injury that may occur before a health hazard is recognized and controlled and the virtually insuperable barriers to recovery for most disease victims within the fault system, it is not surprising that, of those who have taken an interest in this area (nearly all of whom are in the United States), many have advocated the introduction of strict liability or no-fault systems of compensation.[7] There are considerable problems with such reforms, however. One of the justifications for the abandonment of fault is that of 'enterprise liability' that is, that activities such as road use or the use of pharmaceuticals ought to bear the cost of the losses they produce even if these losses are unavoidable (see 6.2–6.3). This means that the loss can be shifted from the victim (effecting the compensation goal) onto the risk creator/controller such as a pharmaceutical company, chemical manufacturer, or polluter through which the loss is spread to those such as employees, shareholders, and consumers who directly benefit from the risk-taking. At a more sophisticated level this goal has been reinterpreted as one seeking to place the cost of personal injuries on the 'cheapest cost avoider' so that deterrence is optimized (see 6.2–6.3). To simplify the following discussion the term 'enterprise liability' will be used to encompass both approaches although it must be remembered that this is a considerable simplification and that they can produce divergent results (see 6.3). What is important at this stage is that in both approaches the

[7] Examples are the current campaign in Britain for strict liability for pharmaceutical damage. See also Y. Cripps, 'Avenues of Compensation for Genetic Engineering Accidents' (1980) 9 *NZULR* 150 and J. Friedman 'Health Hazards Associated with Recombinant D.N.A. Technology: Should Congress Impose Liability Without Fault?' (1978) 51 *South Calif. LR* 1355 (arguing for strict liability for DNA research); M. Franklin 'Tort Liability for Hepatitis: An Analysis and a Proposal' (1972) 24 *Stan. LR* 439 (argues for strict liability for transfusion hepatitis); Pearson argues for strict liability in the areas of medical research, vaccines, and 'exceptional risks'. No-fault schemes have been urged in the US in disease contexts such as toxic waste disease (S. Soble 'A Proposal for the Administrative Compensation of Victims of Toxic Substance Pollution: A Model Act' (1977) 14 *Harv. J. on Leg.* 683; W. Ginsberg and L. Weiss, 'Common Law Liability for Toxic Torts: A Phantom Remedy' (1981) 9 *Hofstra L. R.* 859); asbestos (L. Treiger 'Relief for Asbestos Victims; A Legislative Analysis' (1983) 20 *Harv. J. on Leg.* 1979; E. Anderson *et al.*, 'The Asbestos Health Hazards Compensation Act: A Legislative Solution to a Litigation Crisis' (1983) 10 *J. of Legislation*, 25); agent orange (F. Lalle 'Agent Orange as a Problem of Law and Policy' (1982) 77 *Northwestern Univ. L. R.* 48); pharmaceuticals (*The Times*, 17 Aug. 1982, 9; *Guardian*, 29 Nov. 1982, 3; S. McLean 'Compensation and Medical Injury' (1981) 63 *Scot. L. A. G.* 361); and vaccines (M. Franklin & J. Mais, 'Tort Law and Mass Immunization Programs: Lessons from the Polio and Flu Episodes' (1977) 65 *Calif. L. R.* 754; W. McIntosh 'Liability and Compensation Aspects of Immunisation Injury: A Call for Reform' (1980) 18 *Osgoode Hall L. J.* 584; I. Ladimer 'Legal and Regulatory Perspectives in Mass Immunization Programs' [1976] *Ins. L. J.* 459).

policy decision must be made as to 'what should be a cost of what' activity so that it is known which enterprise should bear which losses.

Some risks are particularly well suited to the introduction of strict liability or a no-fault system because of the relative ease with which, once medical causation has been established, this sort of causal responsibility can be assigned to a clearly identifiable enterprise. An obvious example is the introduction of no-fault schemes for road accidents. Here it is not difficult to define the vehicle event for which benefit is to be payable in such a way as to reflect the crude enterprise liability goals, that is, to internalize costs to the easily definable, responsible activity of motoring. Although pedestrian involvement is an added complication, and targeting the loss to the cheapest cost avoider *within* the enterprise is more problematical (see 6.3), the relative ease of attribution to the enterprise goes to explain the proliferation of road compensation schemes. Similarly, strict liability or no-fault schemes for areas like damage to healthy volunteers for medical research and vaccination are facilitated by the clear relationship of injured to responsible party. It is clear who should bear the risk (both as the responsible risk-taking enterprise and the cheapest cost avoider) and the range of those risks. A comparable approximation can be made in the area of traumatic injuries incurred at work during work hours. The vast majority of such accidents will be sufficiently 'work-related' in both enterprise and deterrence terms to justify shifting the loss to the enterprise (see 6.3), and this provides one explanation for the simplifying gloss courts have often implicitly put on the worker's compensation and industrial injuries schemes in finding a temporal and spatial nexus sufficient to establish work-relatedness for an accident. (The threshold problem for *occupational disease* is one of medical causation not risk allocation—see Chapter 3 and below.) On the other hand, such easy allocation of losses to enterprises is not feasible in the typical non-work disease contexts of disease produced from the use of products and environmental exposures. It is also not feasible in the area termed 'medical misadventure' although for somewhat different reasons. Yet these are areas where proof of fault is particularly difficult leaving many more victims uncompensated and, therefore, areas where reform towards stricter liability or a no-fault system is seen as even more compelling. It is important to understand why the problem of defining the risk covered to reflect enterprise liability is such a problem in these cases.

There are two risk-definition, or 'what is the cost of what', issues here. First, there is the question of overlapping enterprise liabilities—that is, where these policy objectives may point to a number of potentially 'responsible' parties and not give much guidance on how the loss should be apportioned between them. For example, say that it is possible for the victim to prove medically that his cancer was due to asbestos insulation flaking off the ceiling of his residence. On which party or enterprise should the loss fall; the manufacturer of the asbestos product; its installer; or the architect who stipulated its use in the design? In cases of disease from pollution or toxic waste sites, is the correct target enterprise for liability the producer of the waste, the waste disposal agent, or the occupier of the site? In cases of adverse drug reactions should the stricter liability be placed on the prescribing doctor, dispensing chemist, distributor, or manufacturer? To ask which party had the strongest causal link with the disease is to pose a complex question about responsibility and expectations of behaviour, while asking who is the cheapest cost avoider can raise imponderable or highly costly questions, particularly of apportionment. The same point is reached when the second risk definition problem—that of boundary or remoteness issues—is analysed. It may be feasible to ease the first problem by channelling liability to a particular enterprise which, it is estimated, should be the responsible party in most cases. Essentially, product liability reform proposals decree this to be the 'producer' of the product with which the injury was connected (see 5.5). But the question then becomes, for which risks associated with that product is the producer to be liable? Under a policy of enterprise liability it is not generally regarded as just to impose 'but-for' strict liability on a producer for *any* injury to a person arising from the use of his product (for example, every cut from a knife or fall from a ladder) or, say, on a health care provider for *all* deteriorations due to treatment. Unlike the case with vehicle accidents, there cannot be the same confidence that the manufacturing enterprise or medical treatment should be held responsible under enterprise liability. Nor can such broad liability be justified in deterrence terms. The scope of the liability must be drawn more narrowly than this simple 'but for' relation to the product, but on what basis? This is the crucial issue for strict liability reforms. As with the first 'apportionment' aspect of risk definition, the question seems to reduce to one of 'legal attribution' and therefore to subjective notions of responsibility. Ironically, then, the distinction between risks to be ascribed to the enterprise and those which are not, still falls to be decided in complex and costly case by case assessments of attribution/responsibility, etc. comparable to those which the shift to strict liability was designed to avoid. Even in a limited no-fault system comparable problems must arise because it is still necessary to assess who is to be responsible to contribute to the fund and what injuries are covered by that fund (see 5.7 and 6.8).

5.4 *Advance Listing and Elective Solutions*

Some commentators have thought that these costs associated with risk allocation were manageable, but others have sought to ameliorate the risk definition problems which arise upon the abandonment of fault by the advance listing of events to which the new system, be it strict liability or no-fault, would be restricted.[8] This device does not avoid proximity issues, it simply puts limits, and typically very severe limits, on the events for which allocation of responsibility to, say, medical intervention can be forecast with reasonable certainty thereby avoiding costly case by case analysis of responsibility in the listed situations. Given the difficulties in risk definition discussed earlier, and the particular medical causal apportionment problems of disease, it is not surprising that these lists are quite short and concentrate almost exclusively on traumatic injuries such as post-operative complications. They have virtually no application for the victim of disease. This is not unexpected because to retain the virtues of simplicity and low cost adjudication the events must be drawn with specific temporal and spatial boundaries. Earlier the similar approximation in the work accident area was mentioned—that a temporal and spatial nexus was often implicitly accepted as sufficient to establish work-relatedness—and once again it can be stressed that the comparable analogy is not possible for occupational diseases because they are

[8] Respectively, J. Jolowicz, 'Liability for Accidents' 1968) 26 *Camb. J. J.* 50, rejected by Pearson, para. 1648; J. Henderson, 'The Boundary Problems of Enterprise Liability' (1982) 41 *Maryland L. Rev.* 659; and C. Havighurst, 'Medical Adversity Insurance: Has Its Time Come?' [1975] *Duke L. J.* 1233.

not contracted at a recognizable instant of time. Indeed, where a spatial and temporal nexus has been accepted as sufficient in the disease area—the heart cases—a hopelessly irrational system has resulted (see Chapter 3).

Listing of disease risks is not impossible, however. In fact the most well-known example of this cost-saving listing device is the scheduling of diseases for which no-fault workmen's compensation (and later industrial injuries) benefits would be payable. As described in Chapter 3 the list is limited to cases where, on the basis of known medical causes, there is a high probability that the disease was contracted from work conditions. While this system has the minor disadvantage of including the occasional case where, for example, a nurse's tuberculosis was not contracted due to her work, its major drawback is that it excludes so many *bona fide* cases of work-related disease. In fact it is these resultant anomalies which are the principal objection to any system of exclusive listing. This is an especially serious flaw for disease victims since—except in the work context—it is difficult to envisage any area where it is feasible to schedule diseases in such a way as to ensure a high probability of both medical causation and correct legal attribution while maintaining an adequate width of coverage across *bona fide* claims.

Another approach suggested by Professor Jeffrey O'Connell by which fault could be replaced is by allowing one of the relevant parties to elect this course in advance.[9] One form of this idea is elective *third* party no-fault insurance where the potential defendant pre-selects risks for which he would agree to pay the victim's economic losses (less collateral sources) in return for exemption from liability in negligence including liability for pain and suffering. The idea is flawed in a number of respects; for example, potential defendants are likely to elect only risks for which they would have had to pay tort damages in practice. Few disease risks would come within this class. Under elective *first* party no-fault insurance the potential victim insures himself on a no-fault basis for his economic losses due to undefined personal injuries, assigning to the insurer any tort claim he might have arising out of such injuries. The scheme bypasses the risk definition problems of the first suggested scheme but suffers both from the general

[9] J. O'Connell, 'Offers that Can't be Refused' (1982) 77 *Northwestern Univ. L. R.* 589 and refs. cited in footnotes 9, 56, 57 therein.

objection that it retains as an integral feature the lure of lucrative fault-based tort damages, and from the specific problem that it is unlikely to interest insurers especially in the disease context where few cases of disability can be traced to an identifiable tortfeasor. The third variant of these elective solutions is to allow, by statute, defendants to foreclose any personal injuries claim by promptly offering to pay the claimant's economic losses (less collateral sources), the incentive being the loss of the contributory negligence defence and reversal of the onus of proof of liability if the offer is not made. The principal objection to this idea is that it simply facilitates prompt payment to those few already likely to win at least some tort damages (extremely few of whom are disease victims) while abandoning the major aim of no-fault reform, namely the provision of benefits to a broader class of victim.

5.5 *Product Liability Proposals*

Of more potential importance for disease victims are proposals for stricter *products* liability which form the spearhead of the push for reform of tort law in this country. Recommodations along this line have been made by the Pearson Commission, the English and Scottish Law Commissions, the Council of Europe, and the Ontario Law Reform Commission, although from mid-1985 attention has centred on the EEC compromise Directive on the issue.[10] Products liability is a particularly important area for disease victims because most man-made diseases—and virtually all those for which medical causation can be established—are traceable to a 'product' as defined within these proposals. The proposals simplify the risk definition problem of apportionment between potential defendants by channelling liability to the 'producer' of the goods, but the problem of which risks should be visited on the producer remains. In a search to find rational and fair criteria by which liability can be restricted to reflect enterprise liability goals,

[10] Pearson Commission; The Law Commission and the Scottish Law Commission, *Liability for Defective Products* (June 1977) Cmnd. 6831, hereafter 'the Law Commissions'; *Strasbourg Convention on Products Liability in Regard to Personal Injury and Death*, Council of Europe, 27.1.77, hereafter the 'CE Convention'; Ontario Law Reform Commission, *Report on Product Liability* (Ministry of the Attorney General, 1979), hereafter the 'Ontario LRC'; Council of the European Communities, Council Directive on Liability for Defective Products' 25.7.85, hereafter the 'EEC Directive', *Official Journal of the European Communities*, 7 Aug. 1985, L210/29.

reformers have accepted the notion that a producer will be liable for injury due to a product 'defect'. This concept is not difficult to operate when the article contains a manufacturing error to which the injury can be traced, but it proves difficult to operate where the danger is intrinsic in the design (including the presentation of the product such as obscure or inadequate directions for use). The question arises: what is 'defective design' such that liability should be imposed? What is the standard against which such products should be judged? As most man-made diseases for which medical causation could be established arise from articles and substances as designed, the basis on which this question is resolved is crucial to evaluating the significance of product liability proposals for the disease victim.

The consensus which has emerged is that the determinant of 'defective design' should be 'the safety which a person is entitled to expect, taking all circumstances into account',[11] and this is generally regarded to be determinable according to the relative costs and benefits of the product. But there are many uncertainties here crucial to disease cases which reformers fail to address. There is no comment, for example, on the common-sense result that if one concentrates on the idea of 'expectations' then it is hard to see any cases where it would be within expectations of safety for a product as designed to cause disease—unlike many objects which have inherent risks of trauma such as cars and knives. What is more, even if one takes the class of products where toxicity *is* the quid pro quo of efficacy—drugs—it is not known on what basis one should assess the product's costs and benefits. Reformers are silent on the vital question of whether the issue is to be viewed as one of the costs and benefits to the victim or to the public in general. If the former approach is taken the enterprise would escape liability for risks such as baldness due to cancer treatment drugs because even to the victim the benefit outweighs the cost, while if the latter approach is adopted it would also escape liability for risks such as that of internal bleeding from aspirin or adverse reactions to vaccines since the benefit to society seems outweighed by the admittedly high cost to the victim. In other words, reform proposals are silent on the most critical question for disease plaintiffs: how a definition based on expectations of safety would

[11] EEC Directive, Article 6.

handle the question of susceptibility, in particular in the typical disease situation where the victim is unaware of his susceptibility before using the product and the danger is not obvious nor screenable. Similarly, we do not know how far the defendant will be able to rely on warnings of possible health hazards to lower the relevant expectations of safety, especially when pre-use screening for susceptibility to the risk is not feasible.

However, the most confused aspect of the reform contained in the 1985 EEC Directive is that, despite its rejection by the Pearson Commission and Law Commissions, it incorporates the defence with respect to 'development risks' on which the current British government had insisted. This defence applies where a producer can prove that the state of scientific and technological knowledge at the time the producer put the product into circulation was not such as to enable the existence of the defect to be discovered.[12] Despite years of dispute about whether or not to have such a defence, it is a sad and ironic comment on the quality of the product liability debate that even now we do not know quite what this defence means. Clearly it will cover a 'defect' constituted by a risk which at the relevant time was undiscoverable by any scientific means then available. An example would be a health risk associated with use of a drug which is only a risk which affects humans: there is no way to detect the risk before the drug is circulated for human use. Needless to say, such risks are rare. The point is that most risks will have been discoverable by some means. With thalidomide, once the tragic effect on human foetuses was discovered, testing on pregnant animals began, but it took some time to discover a type of animal which showed the teratogenic effect. Will this be relevant? Will a drug manufacturer be able to argue that, even if it had carried out exhaustive tests, the risk would not have been discovered because—given the state of the art at the time—no one would have thought to test the drug on this particular animal? If this line of defence is allowed, what is to prevent its development to cover any risk which was not reasonably discoverable, that is, to prevent the liability narrowing to a negligence-like regime? One might be tempted to dismiss this possible interpretation out of hand for being clearly out of step with the strict liability basis of the reform—a reform which was

[12] EEC Directive, Article 7.

largely motivated by indignation that a negligence regime provides no remedy to such 'guinea pigs'[13] of the process of research and development—except for a fact of political reality. If producers do not seek to have this vitally important line of attack, then the defence would have virtually no scope to operate and the vigour with which they and their supporters in government have insisted on its inclusion appears pointless. But whichever interpretation prevails the point to be made here is that the crucial idea of 'scientific discoverability' is much less clear-cut than reformers assume, and, as a consequence, any assessment of the width of the new liability in practical terms is at best highly speculative.

There are also other flaws associated with the 'development risk' defence. One obvious one is that it creates an inexplicable distinction between defects constituted by risks which are undiscoverable because of the state of the art and risks which are known but unavoidable given the state of the art such as adverse reactions to whooping cough vaccine. Another is that it exacerbates the costly and, for victims of latent disease, the extremely difficult issue of ascertaining the level of knowledge at a time considerably in the past. Yet another is that we do not know what effect post-circulation advances in the state of the art of controlling known risk will have. Does 'presentation of the product' involve any post-circulation warnings? Moreover, quite apart from these major areas of doubt about what the new liability would mean in practice, there are residual and severe limitations on the remedy which would clearly affect disease victims. Principal among these are:

* The proposed time-limits would truncate strict liability in just those cases—long latent diseases—where proof of negligence is most difficult.
* If the liability only arises once products have been put into circulation, it would not apply to classes of victim who are often those most at risk from products hazardous to health such as the producers' own work-force, including experimental scientists.
* The requirement of putting the product into commercial circulation may preclude many environment disease cases such as the poisoning of a nearby community by industrial effluent or other forms of toxic waste.

[13] See e.g. CE Convention Explanatory Report, para. 40. See also Ontario LRC 90; Law Commissions, para. 105; Pearson, para. 1259.

* Medical causation remains the barrier it is at common law as does the requirement to identify the appropriate defendant.
* Parallel issues of apportionability still remain.
* The anomalous situation by which the hypothetical fault of the victim may support a causal argument against the imposition of liability also still remains, it appears.[14]

To sum up: the notion of stricter products liability appears on its face to have advantages for many victims of disease over the existing fault regime. However, specific reform proposals reveal many problems. Some are problems intrinsic to disease cases such as causation and apportionability. But, more importantly, some are due to a continuing and general insensitivity to disease issues such as how to handle the factor of the victim's susceptibility—a phenomenon which is, ironically, particularly striking in reports such as those of the Law Commissions and the Pearson Commission which were to some extent provoked by the non-traumatic injuries suffered by the thalidomide children. The resultant uncertainty combined with the couching of the 'defect' definition in negligence-like concepts of costs and benefits, plus the inclusion of a defence covering development risks, both poorly defined concepts, suggest there would be little if any practical gain to the class of disease victims by these products liability reforms. What is more, to the extent that the reform succeeds it can be attacked for generating the sort of unattractive anomalies inevitably produced by an improved system of compensation (be it tort or non-tort) which is limited as to cause. This latter point is taken up in more detail in following sections where the current vogue for *ad hoc* solutions for limited classes of victim is assessed in terms of its acceptability and feasibility as a way of ensuring better compensation for victims of disease.

5.6 *Non-Legal Barriers to the Initiation and Success of Disease Claims in Tort*

The discussion so far in this and previous chapters has revealed how conceptual and evidentiary barriers prevent all but a tiny fraction of those disabled by man-made diseases having a good claim in tort. Yet of these lucky few even fewer will actually succeed in initiating and pursuing the claim because of a series of

[14] For example, the Law Commission, para. 108, uncritically adopts this common law argument.

non-legal barriers. Some of these simply mirror the practical difficulties faced by accident victims about which there has recently been considerable interest. But in the disease context certain factors take on a particularly formidable aspect. Firstly, there are particular problems in the recognition of medical causation let alone its subsequent proof and link to the defendant. Secondly, there may be particular psychological barriers to the disease victim bringing a common law claim. Little work has been done on this aspect of the perception of disease so that point is speculative but it is possible that, for example, even where external man-made causes are evident or suspected there is a different perception of disease. Chronic disease certainly has a lower public profile and status than traumatic injuries and acute illness.[15] It may be that when a disability cannot be traced to a trauma isolated in time and place from which injury flows immediately and palpably, the victim is less inclined to seek external causes to blame but to see the disease more as the malfunctioning or susceptibility of his own body. Certainly one reason workers give to explain why a common law claim was not brought promptly after diagnosis of the work-relatedness of even potentially fatal diseases is that it was simply 'a risk of the job' (see 2.3). Even when the victim attributes responsibility to a party, it has been found in accident cases that this does not necessarily correlate with a belief that that party ought to pay compensation, the latter being more likely to be a function of the likelihood of a successful claim.[16] If this is so, then the rarity of successful common law disease claims may be having a considerable 'feedback' effect on the number of claims initiated by disease victims. Moreover, the inexperience of lawyers in the underlitigated disease area, together with the greater uncertainty of outcome in many cases may lead to excessively discouraging legal advice. The desire to preserve an existing relationship is also a relevant factor in explaining the failure of accident victims to make and pursue tort claims.[17] In the disease context where symptoms appear only slowly and where there is no overt trauma to alienate the victim

[15] See e.g. F. Peroni, 'The Status of Chronic Illness' (1981) 15 *Social Pol. and Admin.* 43 (in health service priorities); *Thompson* v. *Smiths Shiprepairers Ltd.* [1984] 1 All ER 881, 893.

[16] S. Lloyd-Bostock, 'Common Sense Morality and Accident Compensation' [1980] *Insur. L. J.* 331, 338, 344.

[17] Ibid. 342–3.

from a continuing relationship with the potential defendant, litigation may appear to be an even more hostile act than when it follows a traumatic injury.

Organizational expertise and support would assist disease victims initiating and pursuing claims. By far the most important of such groups are the trade unions whose compensation record at least in the accident field has been seen as a critical recruitment lure. But the unions' track record in the disease area has been poor with most interest centering on the expansion of the industrial injuries schedule of diseases.[18] Although there has been more interest in recent years, the latency of many diseases means that many workers will only first experience symptoms of disease after leaving the relevant place of employment and perhaps the union itself, making it more difficult to obtain union advice and support. In the non-occupational disease context there is no general institutional support, as there is, say, for road accident victims who are compulsorily insured.

Even when the victim is keen to sue, there are particular difficulties associated with the potential defendant in disease cases. Even if identification of the relevant defendant is possible, there are a number of special problems in getting the disease defendant into court. The long latency of many diseases means that between exposure and the victim becoming aware of his right to sue the responsible party may be untraceable or have gone out of business. This is a particular problem, for example, in cases of disease from toxic waste disposal sites. It is also an acute problem in contracting industries, a well-known example here being the Welsh slate-quarrying industry where this problem generated significant and successful political pressure, precipitating a special statutory compensation scheme. Disease claims themselves may contribute to this phenomenon; for example, in the more litigious context of the United States a number of asbestos manufacturers have sought the protection of the bankruptcy laws from the flood of disease claims for which inadequate insurance had been taken out during the exposure period. The problem here with long latent disease may not simply be this one of underinsurance: in some cases insurance records have been lost so that the relevant insurer cannot be identified.

[18] e.g. inactivity on industrial deafness: *Thompson* v. *Smiths Shiprepairers Ltd.* [1984] 1 All ER 881, 892.

Even if the defendant is identified and backed by adequate insurance, there may be other barriers to pursuing a claim to court. For example, while defendants (in practice their insurers) are relatively risk neutral,[19] there are considerable pressures on a plaintiff to settle especially where, as in disease cases, the uncertainty of outcome is high. Moreover because of the duplicative nature of many man-made disease cases, defendants and their insurers have greater incentive to resist claims and co-operate together both on medical causation and fault issues. But there are also important pressures on the defence to settle. It is less expensive than litigation and, of special note in disease cases, it avoids the bad publicity of a court judgment while preventing crucial issues such as medical causation being aired and judged. The role of common law claims in publicizing hazards is especially important in the disease area where injury is inflicted gradually and without the sort of dramatic event which might focus the attention and sympathy of the public. Settlements in these cases have the effect of inhibiting awareness both of the hazard and the possible legal liability of the potential defendant, thereby inhibiting new claims. Of course critics of the tort system do not suggest that all cases should be litigated. The courts would not be able to handle the case-load. The problem is that, although a necessary part of tort law in practice, the settlement process undermines the compensatory goals of that law by its discount for uncertainty of outcome and by the stifling role it has on the publicizing of little-known areas of hazard and liability. On both scores the settlement process operates particularly harshly towards disease victims.

5.7 *Abandonment of Tort: Limited Schemes*

The past two decades have seen widespread dissatisfaction with the way tort law handles personal injuries. Some of this concern centres on how well tort deters dangerous conduct (see 6.3–6.7), but the principal debate in England arises from the widespread acknowledgement that in substantial measure it fails to achieve its 'corrective justice' goal, particularly its compensatory aspect. The tort system is attacked on the basis of its delay, inefficient and costly administration, but in particular on its unpredictability of outcome both on the question of liability and on the adequacy of a

[19] J. Phillips and K. Hawkins, 'Some Economic Aspects of the Settlement Process: A Study of Personal Injury Claims' (1976) 39 *Mod. L. R.* 497, 504.

once and for all lump sum payment. Moreover, as the preceding section illustrated, there is an important gap between liability in theory and outcomes in practice, since the settlement process rewards negotiation advantages not recognized as relevant in tort theory. Appreciation of these failures of the system to compensate those few it purported to serve (that is, victims of fault) was accompanied by doubts as to whether the system could adequately be reformed to meet its limited compensatory goal. But more important still was the fact that analysis of the weaknesses of tort in providing compensation to these few led to a re-evaluation of the fault principle itself—the moral basis of negligence liability—towards a pure compensatory goal (see 6.2).[20] The focus shifted from the relative justice of the relationship between plaintiff and defendant to that between victim and victim. Once this step was accepted, the 'failure' of tort to compensate appeared far more serious since the class of people denied compensation flew open to include virtually all accident victims. Henceforth, the preference negligence law gave in compensating a very small percentage of people who were fortuitously victims of 'fault', while the bulk of victims went without damages was seen, a little elliptically, as a fatal failing of the fault system.[21] Originally concern had centred on road accident victims where the fault criterion of recovery was palpably artificial in most cases. But the principles upon which fault was attacked and abandoned were immediately applied to other accident victims and ultimately, with considerably less emphasis, to victims of disease.

The contemporary focus on the compensation goal together with a rejection of the tort system as its vehicle has led to a number of non-tort no-fault systems for compensating particular groups of victims. Schemes for road accidents predominate (along with those for work accidents which originated from an earlier historical period of dissatisfaction with the fault system)—no doubt largely because of the ease with which schemes can be defined and financed from systems of compulsory insurance (see 5.3). But if

[20] Atiyah, *ACL*; P. Atiyah, 'No-Fault Compensation: A Question that Will Not Go Away', 280–5; I. Englard, 'The System Builders: A Critical Appraisal of Modern Tort Theory' (1980) 9 *J. Legal Studs*. 27–8, 68; A. Tunc, 'Tort Law and the Moral Law' (1972) 30, *Camb. L. J.* 247; B. Hepple and M. Matthews, *Tort: Cases and Materials*, 2nd ed. (London, Butterworths, 1980), 7.

[21] See e.g. Fleming, op. cit., 316; Palmer, *Compensation for Incapacity* (Wellington, OUP, 1979), 23, 35.

tort is to be rejected for its failure to compensate the victims of accidents, its abandonment in the case of disease is even more urgent given the rarity of recovery of damages for disease. Consequently, one of the results of recent concern with victims of man-made disease has been that in isolated situations limited non-tort systems of compensation have been set up both in Britain and overseas. These range from private *ex gratia* funds paid out by drug companies following adverse drug reactions (for example, the thalidomide trusts) or by employers to employees suffering a particular occupational disease (such as the schemes set up by the National Coal Board and British Nuclear Fuels) to state schemes limited to a particular hazard such as vaccines or a particular group of disease victims. The principal criticism to be made of these schemes concerns the anomalies of treatment within the class of disabled which they create: a striking example of this from the United States is the Black Lung programme which has involved 500,000 miners and dependants at a cost of $1 billion (US) a year—40 per cent of total workers' compensation payments—to 1/300 of the working population.[22] Before discussing the justification advanced for such limited schemes, it is worth examining just how successful they are in operation for the compensation of certain man-made diseases.

Representative of the first type of scheme is that operated by ICI to compensate victims of its heart drug Eraldin (practolol) which was marketed in June 1970 and voluntarily withdrawn in October 1975. The scheme has been criticized in the press both for indirectly limiting publicity about this 'forgotten drug disaster'—a result which, of course, effective compensation schemes are likely to produce—but more significantly for raising expectations of compensation among claimants who are ultimately unable to satisfy ICI of the requisite level of causal proof. ICI accepts only some of the recognized adverse reactions to the drug as compensatable under its scheme: and of these the most common reaction (dry eye) is also a 'normal' condition in people of the age-group which took Eraldin (average age, 62) so that proof of causation is particularly difficult: in the first five years of the scheme, 1,100 of the

[22] See J. Strader and P. Sheehe, 'Federal Black Lung: Ten Years of Legislation and Litigation' (1981) 16 *Forum*, 525, 550 n. 98; K. Morgan, 'On Disability, Demagoguery and Dialectic' (1979) 14 *Forum*, 473, 477; Anon., 'Compensating Victims of Occupational Disease' (1980) 93 *Harv. L. R.* 916, 929.

2,600 who made formal application for compensation were rejected outright.[23] The problem of medical causation produces an even higher rejection rate for claims under the government's limited vaccine damage payments scheme. Vaccines, particularly that for whooping cough, can cause convulsions in children and lead to brain damage but convulsions can also occur 'naturally' in the relevant age-group (3 to 8 months), so that satisfying the statutory balance of probability requirement has proved extremely difficult[24] Medical causation is not the only problem with these limited schemes—there are also the risk definition problems of what is and what is not to be covered by a limited scheme. A good example of the difficulties which straddles both traumatic and non-traumatic injuries is recovery for 'medical misadventure' under the New Zealand scheme.[25] Even if the practical problem of distinguishing the relevant disability from the underlying condition can be overcome, and it is established that the deterioration would not have happened but for the treatment, there remains the question of whether the loss is within the range of normally expected medical risk or failure (which the patient is assumed to have accepted in return for the potential benefits of treatment). Some cases are clear-cut. Thus, infection from an unsterile surgical instrument constitutes medical misadventure while relatively mild adverse drug reactions go uncompensated. But many cases are not so clear and the case by case assessment of these has been a major and costly problem under the scheme.

In short, then, even when both the fault and tort systems are abandoned there are still at least two intrinsic and costly problems in operating compensation schemes covering disease which are defined in terms of cause: medical causation and remoteness. One may summarize other problems too; the drawbacks to a once and for all lump sum benefit; a loss of individualization; non-pecuniary losses may not be covered; the adequacy of benefit is less open to public scrutiny especially in private schemes; and crucial issues

[23] *Guardian*, 3 Mar. 1981, 8; A. Melville and C. Johnson, *Cured to Death: The Effects of Prescription Drugs* (London, Secker & Warburg, 1982), 147.

[24] Parliamentary Debates HC Session 1981–2, Written Answer 2 Mar. 1982, col. 124.

[25] Accident Compensation Act 1982 (New Zealand), No. 181, s. 2. See 'Medical Misadventure' [1981] NZACR 244; Palmer, *Compensation for Incapacity*, 255–8. Compare *Re Muir: Decision 561* [1981] NZACR 228, and *Re Lloyd: Decision 737* [1982] NZACR 259.

such as apportionment between causes and the treatment of abnormal susceptibility still present unresolved problems. In any case, the overall practical importance of these schemes in compensating victims of disease is not significant as the number of those who benefit is very small (the exceptions are schemes covering work diseases). But the schemes are significant in representing something more disturbing—a growing tendency to treat compensation issues separately and preferentially according to the medical cause of disability, an apparent illogicality not present in the tort system. The remainder of this chapter examines the justifications and criticisms of these and other types of preferential treatment in compensation systems.

5.8 *Justifications for Preferential Treatment*

At the outset it must be conceded that some *ad hoc* schemes may work well in compensating a class of victim. The thalidomide trust is a good example where benefits are generous, and demarcation and medical cause problems relatively insignificant. The first public and still dominant scheme—workers' compensation for industrial injuries—is also far more successful than tort in delivering compensation to a sizeable class of disease victim. Relative to tort, such schemes are usually both speedy and cheap to administer—for example, many schemes do not require individual identification of the responsible party—while avoiding the negative effects on hazard investigation which litigation may pose. It is not on the basis of efficiency that the *ad hoc* schemes are attacked but on the anomalies they create between the relatively high level of compensation received through them by a lucky few and those who receive no such compensation, even though their disability and need is the same. Unless a rational justification can be advanced for such preferential treatment, it is argued, justice demands that like cases be treated alike[26]—a principle which is usually taken to lead to the recommendation of a comprehensive compensation scheme such as those devised by the Woodhouse Royal Commission (New Zealand) and Committee of Enquiry (Australia). Are there such justifications?

Once the fault basis of liability is abandoned then, whether it is by limited extension of strict liability or a non-tort scheme, there

[26] Atiyah, *ACL* 216.

seems no *general* theoretical rationale for preferential treatment in compensating the disabled. There are various reasons suggested in individual cases why a particular class should be treated differently. Sometimes this just amounts to stressing a particular failing in the operation of fault liability with respect to that class of disabled, but are there *additional* justifications? Perhaps the most well-known (notorious?) collection of *ad hoc* justifications is in the Report of the Pearson Commission. This Commission, widely attacked for its lack of a coherent social policy basis and generally limited view of its terms of reference, has had little effect on the position of victims of man-made diseases and it is convenient at this stage to make a slight digression to explain why.

The Commission's terms of reference were considered by its members not only to exclude 35–40 per cent of accidents but also all non-occupational illness.[27] Why was disease thought to be excluded? Perhaps it was because of the use of the term 'injury' in the terms of reference. But the point is that nowhere is it explained, nor is there an acknowledgement of the definitional conundrum of separating traumatic injury from man-made illness and these from other disability (see 1.2 and 5.9). The Commission regarded 'occupational diseases as implicitly covered because in our social security legislation they are treated as if they were industrial injuries'[28] which seems a somewhat strained and atypically generous distinction to make about the meaning of 'injury' in this *one* context, suggesting some unstated preference for this group of disease victims. But having decided on whatever grounds to exclude non-accidental illness, the Commission then evaluates at least two classes of 'injury' with examples of what could be non-traumatic disability—antenatal 'injury' and product 'injuries'. Indeed, it was non-traumatic pharmaceutical damage in these cases—specifically the thalidomide disaster—which had precipitated the Commission in the first place and it was clearly intended to be covered. This was probably also true of the entire class of non-traumatic man-made diseases since virtually all are attributable to the use of goods or the act of omission of another.

One consequence of the failure to distinguish this class of non-traumatic man-made disability from illness in general is to confuse the data on 'injuries' collected by the Commission (do they include

[27] Pearson, para. 239. [28] Pearson, para. 767.

drug damage?). A more fundamental result of this failure can be appreciated when taken together with the fact that in one area Pearson felt justified in going beyond its construction of the terms of reference to include all disabled of one class: severely handicapped children. The appropriateness of selecting this group is discussed below; what is crucial here is that the inclusion of all disabled from that class—victims of trauma, man-made disease, and 'natural' illness alike—was justified on the basis that here 'it is simply not feasible to devise a separate scheme of compensation for those children who are within our terms of reference [and] . . . it would not be sensible to allow the constraints of our terms of reference to prevent us from seeking the right solution'.[29] But why was this rationale not applied to the whole range of disability since, for example, there are vast areas where it is impossible medically to distinguish disease cases accepted as within the terms of reference (occupational disease, drug damage, etc.) from those outside? Major examples are heart disease, cancer, and bronchitis. Once again the failure to identify the class of man-made disease perpetuates a focus on traumatic injury which distorts the thrust of reform.

The espoused philosophy of Pearson was a shift from tort towards no-fault but because of its limited reading of its terms of reference it felt it could not consider a comprehensive scheme. Thus the 'logic' was to deal separately with conventional categories such as road accidents and assess the feasibility of limited extension of strict liability or the introduction of no-fault compensation. In other words, despite espousing a needs basis and the goal of removing compensation anomalies between the disabled, the whole thrust of the report was to increase the anomalies and complexities by this *ad hoc* approach which at best delays reform and at worst entrenches preferences which will further impede reform. Given this nature of the Pearson Report it is not unexpected that one particular argument advanced in defence of preference formed a recurrent theme of its analysis: that reform should not erode rights already enjoyed, albeit by a tiny fraction of victims—at least if others are going to retain this 'right'.[30] (This is also essentially the modern defence of the industrial preference—see below.) Thus the Commission thought the introduction or

[29] Pearson, para. 1488.

[30] Pearson, paras. 302–6.

acceptance of no-fault recovery in one class, for example road or work injuries, should not be accompanied by abolition of that class's tort remedy as this would disadvantage a subclass (victims of road or work injury due to *fault*) when compared to those victims of fault outside the groups altogether, who retain access to tort level damages. The obvious 'ratchet' effect of such reasoning reflects the inevitable complications and illogicality produced by an *ad hoc* approach concentrating on one class of victim. Here in one breath preference is approved in one area (no-fault) but objection to preference is used to produce a doubly-favoured position in another situation (fault).

Preferential treatment might also be defended in terms of economic and administrative feasibility. The existence of a suitably adaptable distribution of insurance was cited by Pearson as a factor supporting limited extension of strict liability (for example, to products) or no-fault compensation (for example, to roads).[31] Restriction of the strict liability recommended by Pearson for exceptional risks to only statutorily listed risks also suggests an unstated pragmatic approach: only risks where attribution of responsibility is administratively feasible in advance would attract the liability. [As noted earlier, exceptionally few disease risks are suitable for such advance listing—see 5.3.] Groups may also be singled out for preferential treatment because of the number and severity of that class of injury. This was one of the factors cited by Pearson in support of a no-fault scheme for road accidents. Another stated factor here that, unlike industrial injuries, road accidents occurred randomly and were not confined to a particular class, was particularly bizarre since its converse has been a traditional *support* for the preferential treatment of another class (work injuries . . . see below). Moreover, many man-made disease injuries—typically those from pollution and products—are also random in their impact. A third reason in support of preference for road victims was that the activity 'is an essential part of everyday life, of fundamental importance to the economy as a whole and to the mobility of individuals'.[32] Why these factors make road accident victims more worthy of sympathy and generous treatment than victims of risks inherent in other worthy activities of modern society is unclear. The reader is left with the

[31] Pearson, paras. 1641, 1235.

[32] Pearson, para. 996.

suspicion that road injuries, like work injuries, are to be given preferential treatment simply because they are a large visible class of disability for which a limited improved system of compensation is economically, administratively, and politically feasible.

In one area preferential treatment is defended on more philosophical grounds. In certain areas such as military service and volunteers for medical research, clinical drug trials, and vaccination, the individual has undertaken a risk *inter alia* for the benefit of society, often in direct response to official encouragement. The rationale for preferential treatment is, therefore, twofold: volunteers should not be discouraged by the risk of uncompensatable injury; and society should shoulder the risk from which it so directly benefits. While considerably more cogent as a rationale for preferences, there are a number of considerations which detract from the generality of the argument. First, as Dworkin points out,[33] other activities such as the wearing of seat-belts may be strongly recommended or even made compulsory by government concerned to avoid the economic costs of road accidents to society, yet no preference is given to injuries due to seat-belts. Secondly, one might launch an even more potent argument in favour of preference for those who have been the *unwitting* 'guinea pigs' for society by being exposed to new chemical and pharmaceutical preparations. Indeed, this has been used as an argument in support of a regime of products liability which dispenses with any defence based on the state of the art (see 5.5). Similarly it might be asked why the dubious ethical support for the Criminal Injuries Scheme—to salve 'the nation's conscience at its inability to preserve law and order'[34]—does not also apply to the nation's inability to police pollution and other disease hazards.

Finally, there are two relevant groups for which special concern and preference is given: children and workers. Is this justified? Firstly, children: the discussion earlier touched on how the Pearson Commission went outside its perceived terms of reference to look at the plight of all severely handicapped children, using the problem of medical causation and a needs argument as justification. But the Report gives no explanation why these arguments apply only to the case of children. Certainly the post-thalidomide world

[33] G. Dworkin, 'Compensation and Payments for Vaccine Damage' [1978–9] *J. Soc. Welfare L.* 331, 334.

[34] Pearson, para. 1588.

is now highly sensitized to the plight of handicapped children, particularly with congenital defects, so that the notion of coldly distinguishing cases on the basis of causation seems offensive. But irrationally confined sentimentality does not seem an admirable principle on which to base reform. For example, the victims of the drug DES were just as innocent as the thalidomide children but their antenatal drug injury did not manifest itself until their mid to late teens. Indeed, the injury in many cases—vaginal cancer—was more serious than many of the thalidomide injuries. Why do they not also deserve assessment on the basis of need? Why are children special? If a group of disabled are to get preference, adults who have already contributed to society may have a stronger claim to compensation resources. Indeed, it is the special status of workers which has given rise to the most well-known anomaly: the industrial preference. In practical terms this industrial preference finds its principal expression in disablement benefits (for work-caused disability) which have no analogue in the rest of the social security system. [In the Netherlands it results in a unique preferential system of benefits for earners which cover *all* disability not just that caused by work, see 7.7, Option IV.] The traditional arguments in defence of the preference which were accepted by Beveridge are now widely rejected. For example, the argument that those who contribute to the economic well-being of the community deserve special protection when they suffer work-related injury, does not easily or necessarily translate into preferential treatment. As an argument in favour of the *adequacy* of support it may be a potent line of reasoning, but is it a good reason for *better* treatment especially one not geared to the value of the victim's economic input? Indeed, Brown has concluded that the principal argument of unions and others is no longer a defence of the preference *per se* but of its more adequate level of compensation.[35] The Disability Alliance has also stressed that this position is not in conflict with the modern thrust towards reform on a needs basis which could be accomplished by 'levelling up' benefits to the level under the industrial injuries scheme.[36] In short, then, while there may still be arguments in favour of the

[35] J. Brown, *Disability Income Part I: Industrial Injuries* (London, PSI, 1982), 282, 295.

[36] Disability Alliance, *Reforming the Industrial Injuries Scheme: The Wrong Priorities* (London, DA, 1982), 3–4.

separate administration of industrial injuries benefits—for example, causal attribution is important in deterrence efforts (6.3–6.8, 7.10)—there seems to be no modern principled rationale of its preferential benefits.

There is one final and extreme 'justification' which might be given for preferential treatment: political power. Since society unevenly distributes other benefits on the basis of sectional interests, is it objectionable that the division of the compensation pie should be open to the influence of pressure groups and political power? Certainly a number of *ad hoc* schemes have originated in this way and it is widely regarded as the reason for the survival of the industrial preference. Similarly, the media campaign for drug damage compensation and Pearson's concern with severely handicapped children may be a function of the greater political feasibility of getting additional benefits for such poignant victims. It is simply easier to campaign on the hardships of a specific class of victim. The sympathy of the public and hence political power can be engaged more easily on a focused misfortune due to a specific cause. This is especially the case where—as in the case of drug damage—the general public can identify with the injured as innocent victims of a daily risk which the public itself runs. Corporate power is also important in entrenching compensation anomalies: the Manville Corporation has conducted a massive 'well-orchestrated' publicity campaign for a federally organized (and part subsidized) asbestos compensation scheme in the United States.[37] As a justification for preferential treatment this idea of political pressure clearly runs against the current tide of egalitarianism. Such crude 'survival of the fittest' ideas are intellectually distasteful to most participants in this reform debate. Still there are a number of issues in the compensation debate—the earnings related principle, the treatment of illness—which involve value-judgements of a similar ideological flavour and where a rigorous egalitarian approach is often eschewed. In the light of these matters, it is not quite so easy to dismiss the 'political power' justification as outdated, at least in tandem with the notion that such preferences can act as pace-setters for benefits counterbalancing the erosion of benefit levels by cost cutting governments. The point is taken up again in Chapter 7.

[37] Epstein, op. cit.

5.9 *Rejection of Preferential Treatment*

The argument against preferential treatment, be it between victims of fault and others or between beneficiaries of limited non-tort schemes and others, is the relatively simple one, that just distribution of resources requires equality of treatment based on need. In this context the availability of an identifiable tortfeasor (under a fault or strict liability regime) or the cause of the disability (under limited non-tort schemes) ceases to be relevant. Other subsidiary pragmatic arguments against preferences such as wastefulness of demarcation disputes are also advanced but the principal line of attack is the one of distributive justice. It is feasible that some social goal might outweigh this principle. But the preceding examination of 'rationales' for preferential treatment revealed a lack of sound policy on this point at both a general and specific level. The conclusion is that once the tort system is abandoned as an appropriate compensation mechanism there is no rational justification for preferential treatment of the disabled. The attitude is exemplified by Atiyah: 'All the disabled are entitled to equal sympathy and equal support from the State and *ad hoc* treatment of special groups can only lead to the abandonment of all rationality in policy.'[38] In general, the result is a shift towards the welfare philosophy based on needs underlying the social security system. This philosophy is typified by the National Health Service's rule of equal access to treatment and its growing popularity by the introduction of new non-contributory benefits such as mobility allowance, attendance allowance, invalid care allowance, and new invalidity pension. Even the Pearson Commission's eccentric approach to personal injuries espoused, at least in theory, a needs basis and in one limited area it was the excuse used to breach the limits of its perceived terms of reference. The 'needs' philosophy has become so widely accepted as the desirable principle in this field that it has almost become conventional for reformers to advocate a comprehensive Woodhouse-type solution or a National Disability Income to replace the patchy network of tort and non-tort sources of compensation. Moreover, it is arguable that even the recent recession and conservative mood of government will not defuse,

[38] Atiyah, *ACL* 431.

merely delay, this thrust for comprehensive reform. But how does all this translate in its application to disease victims?

Of course arguments to reject fault, tort itself, and *ad hoc* schemes in favour of a needs basis are even more powerful in the area of man-made diseases: proof of fault is more rarely possible than in accidents and in only a few classes of disease victim would the introduction of stricter liability assist. Any strict liability or limited no-fault solution would still be hampered by risk definition and medical causation problems that would exclude most victims of man-made disease. Thus, even in such politically favoured areas as work-related hazards, current schemes do not even attempt to cover all occupational diseases—indeed, it is thought that 80 per cent are excluded from the Industrial Injuries Scheme in practice[39] —nor all victims of scheduled diseases (for example, those who live with asbestos workers and contract asbestos cancers). Moreover, this pattern is repeated in other jurisdictions, for example, in the United States it is estimated that only one-third of deaths and 1/13 of disablements caused by work disease result in any workers' compensation award, while in Canada the coverage of workers' compensation for occupational disease is so narrow in practice because of problems of medical causation that this, rather than the corresponding failure of tort to deliver compensation, is used by one commentator to argue for a comprehensive compensation system.[40] In short, man-made diseases are particularly effective in highlighting both the inequity of compensation schemes based on cause and their substantial unworkability in many cases. It would, therefore, seem logical that at least one touchstone by which any proposed reform based on needs should be judged would be how it treated these victims of man-made disease, paradigm examples of the failure of tort and limited non-tort schemes. More importantly, the replacement of fault with need as the criterion of compensation policy inevitably leads to the conclusion that all disease victims, including those due to, say, genetic factors, should also be given equal treatment. Yet the pace-setting New Zealand scheme excludes all but occupational disease. Even the Woodhouse

[39] See Pearson, vol. 2, para. 173.

[40] See respectively Anon., 'Compensating Victims of Occupational Disease' (1980) 93 *Harv. L. R.* 916, 925; P. Weiler, *Protecting the Worker from Disability: Challenges for the Eighties* (Report submitted to Russell Ramsay, Minister of Labour, Apr. 1983), 10–11, 73.

(Australia) Report considered treating disability due to accidents with preference, at least in terms of speed of reform. In other words, even reformers who adhere to a needs basis often find it necessary, usually on ostensibly pragmatic grounds, to draw a distinction between two classes of disabled. The interesting fact is that this is drawn between injury by accident (class 1 . . . Figure 1.1) and victims of non-traumatic illness (classes 2 and 3). Why is this so? If particular sympathy is felt for accident victims because somehow this is the price for living in a highly complex industrial society then that applies equally well to the man-made diseases of class 2. Even tort law sought to give some people disabled in this way a remedy.

Of course the reason why a distinction is not made on the basis of disability caused by man-made versus 'natural' causes is because of the intractable problem of medical causation. It is known, for example, that 80 per cent of cancers are due to environmental (societal) factors, factors over which control can be exercised, but it is usually impossible to diagnose origin in individual cases. It is also known that death rates vary between regions because of the inadequacy of medical facilities—that over 20,000 deaths a year in England and Wales are avoidable by reasonably attainable levels of treatment[41]—but again both medical and policy attribution would be difficult, and no doubt controversial, in any particular case (see 1.2). Indeed most types of disease straddle the theoretical line with some, usually the minority of, cases being *positively* attributable to man-made causes and others for which this is not possible. To attempt to define a compensation system which covers classes 1 and 2 but not 3 would simply be too costly to administer and raise too many controversial issues while exposing the reform to the cry of 'anomaly' as some, say, cancer victims received benefit while others did not. In contrast, the line between classes 1 and 2 is relatively workable with problem areas concentrated on issues such as the 'process' distinction and medical treatment injuries. It is, therefore, for solid practical reasons that a distinction *within* the disease category is not adopted. Why a line should then be drawn between injury and all

[41] J. Charlton *et al.*, 'Geographical Variation in Mortality from Conditions Amenable to Medical Intervention in England and Wales', *Lancet*, 26 Mar. 1983, 691; *The Times*, 27 Mar. 1983, 4.

disease—whether it is simply a pragmatic compromise on the path of reform or belies deeper perceptions as to the aim of law reform in this area—is critical to the future direction of the reform debate concerning compensation for personal injuries. It is explored further in Chapter 7.

6

Deterrence and Health Hazards

6.1 *Introduction*

The focus of the preceding chapters has been the conceptual and practical problems faced by victims of man-made diseases when they seek a remedy in tort, although some mention has been made of non-tort systems of compensation to illustrate the persistent nature of some of these problems. The conclusion has been that tort has not provided, nor could it be made to provide, an adequate remedy in practice for this class of personal injury victim. Little attention has been paid to the legal theories put forward in defence of the tort system. If a system cannot be made to operate adequately in practice—and causal barriers alone eliminate most man-made disease claims—it seems somewhat futile to analyse its benefits in theory. However, the touchstone of the preceding chapters has been the investigation of situations and concepts given a new light by the issues raised by a disease perspective. This and the following chapter, therefore, turn to two particular areas of legal theory where this is the case: the debate about the economic role of 'compensation law'; and the fundamental policy bases of compensation reform.

6.2 *Justifications for Tort Law in the Area of Personal Injuries*

There is no consensus about what the modern law of tort is or should be aiming for in providing an action for damages for personal injuries. It is well accepted that if the principal goal were the compensation of victims, tort should be replaced by some sort of collective comprehensive system of compensation. But tort law, by demanding a one-to-one coextensive relation of compensation paid to a victim with the amount extracted from a particular and identified source, produces a number of other effects besides compensation and some have argued that these are worth preserving, even worth giving primacy to. If these arguments prove persuasive, inroads into, or supersession of, the compensatory goal might then be accepted.

Historically, the moral goals of tort law have been taken as paramount in this sense. The idea was one of 'corrective justice' as between two individuals, the blameworthy party being made to atone by compensating the victim whose injury he has caused. Tort scholars are familiar with the current arguments for and against the fault basis of negligence law. There is, of course, nothing irrational in a system which predicates compensation of victims on the finding of fault in an identifiable injurer, leaving other victims to fend for themselves. But modern critics illustrate how crudely, if at all, this moral basis of negligence law operates by pointing to internal inconsistencies in the system like vicarious liability, the objective standard of care, and the lack of correlation between level of fault and level of damages, and to external realities such as liability insurance premiums which are inadequately rated to reimpose representative losses on defendants. In the United States ethical justifications supporting a system of stricter tort liability have arisen to rival theories based on economic efficiency,[1] but there seems little interest in Britain in trying to resuscitate moral justifications in this area of the law, presumably because of the current ascendancy of the compensatory goal (see 5.7 and 5.9). In any case, an appreciation of the particular issues raised by disease cases does not add anything new to this debate except that, by highlighting the practical and conceptual barriers to the imposition of liability on wrongdoers, it provides further evidence of the failure of the 'fault' system to operate in practice. Similarly disease cases add nothing to the associated debate on the usefulness or appropriateness of tort law as a social safety value enabling injured parties to vent their indignation—even vengeance—against those responsible.

One role of personal injury litigation which is of particular value in disease cases is in terms of publicity. In this way insidious hazards may be exposed, generating not only awareness among the public at large but also awareness and action in legislative and regulatory bodies. If tort were to be abolished in the health hazard area therefore, it would be especially important to institute a mechanism at least as efficient—hopefully more efficient—at

[1] R. Epstein, 'A Theory of Strict Liability' (1973) 2 *J. Legal Stud.* 151; G. Fletcher, 'Fairness and Utility in Tort Law' (1972) 85 *Harv. L. R.* 537. On these see G. Schwartz, 'The Vitality of Negligence and the Ethics of Strict Liability' (1981) 15 *Ga. L. R.* 963.

focusing such awareness. The usefulness of litigation publicity in deterring the individual wrongdoer is less clear. First, it is a particularly crude instrument because the deterrent can apply with almost equal force even in cases where liability is not ultimately established and because litigation does not expose to publicity defendants who are most obviously in the wrong since these will usually settle out of court. Moreover, unfavourable publicity in the absence of litigation can be just as potent a deterrent and, in any case, public relations considerations tend to produce irrational spending patterns on prevention.[2] In short, arguments in defence of tort based on the deterrent potential of its attendant publicity seem ultimately unconvincing.

Occasionally tort law is rationalized in terms of its ability to spread losses away from a concentrated loss on the victim. It does this by shifting the loss onto defendants who are almost invariably insured so that the loss is ultimately borne by those who benefit directly from the defendant's activity, for example, its shareholders, employees, and customers. While it cannot be doubted that loss spreading is an effect of tort law, to advocate it as a major goal makes no sense, since it cannot explain or justify the selection of defendant and distribution of liability rules. Indeed, taken to its logical conclusion it would—like its twin goal, compensation—select as loss bearer not an individual tortfeasor but the entire community which is both the best loss spreader and has the deepest pocket. Often what is really intended by 'loss spreading' is a more limited form of loss distribution which has a much more rational basis: the allocation of costs to activities or enterprises to which the personal injury is attributable, and through these to those who have benefited from the risk taking. It was previously noted (5.3) that this is often one of the cruder justifications for stricter liability, but that since it depends on legal attribution of harm which in turn hinges on rather indefinite notions of responsibility, its rational formulation in either ethical or economic terms can be particularly difficult in individual cases.

Considerably more focus has been given to this risk/loss distribution approach by modern economic theory. Rather than sharpening the definition of who the defendant should be by developing the notion of cause and responsibility, economists—

[2] R. Pierce, 'Encouraging Safety: The Limits of Tort Law and Government Regulation' (1980) 33 *Vand. L. R.* 1281, 1288.

mostly in the United States—have asked what liability rules should be applied if priority is given to the economic goal of efficiency both in terms of achieving the optimum level of accidents and the improvement of resource allocation. [Since resources are limited avoidance of all personal injury is not a rational goal: the economic optimum is to minimize net social costs (that is, injury costs plus prevention costs).] So, for example, a polluter may be 'at fault' and have 'caused' the victim's injury, yet it may be preferable to leave the loss on the victim because discouragement of living in the neighbourhood may be the more efficient solution overall. It is this theory of tort law which warrants closer examination in the context of this book, not only because of its ascendance in the United States but because the disease case is perhaps the best illustration of the theory's intrinsic weaknesses and because the questions raised here have equal application to the issue of how non-tort compensation funds should be financed.

6.3 *Normative Economic Theory: Its Two Limbs*

Economic theory emphasizes the fact that tort law can have useful incentive effects towards the prevention and deterrence of future personal injuries. The central notion is 'market deterrence' although subsumed under this term are two distinct limbs, one concerned with the preventability of the risk and the second concerned with deterrence via the price mechanism.[3] Failure to distinguish these limbs can result in significant confusion and error in applying the theory, for example where commentators seeking to promote deterrence concentrate on the preventability of the hazard, ignoring the advantages of price deterrence in the reduction of injury from unpreventable/undiscoverable risks. It is, therefore, intended to refer to the first limb which focuses on the conscious individual response to the burden of liability as 'prevention' while describing the second limb which deals with the market's response to the allocation of costs via the price mechanism as 'price deterrence'.

The first, 'prevention' limb of the theory argues that if the cost of personal injuries is placed on the 'cheapest cost avoider' he will have an incentive to prevent similar future losses by taking cost-justified precautions. Much debate has taken place about which

[3] G. Calabresi, *The Costs of Accidents* (New Haven, Yale University Press, 1970), 73; Pierce, 'Encouraging Safety' 1289–90.

liability rule, fault or strict liability, is more efficient in this respect. At first it may seem that the level of prevention incentive ought to be the same.[4] Under a fault rule, it is argued, the defendant should only be held liable if the loss caused by his conduct exceeds the cost to the defendant of avoiding the damage. It only imposes (internalizes) on the defendant the cost of such 'cost unjustified risks' and it is only these which the rational defendant is, therefore, motivated to take precautions against. On the other hand, under the purest form of a strict liability regime—that is, 'full strict liability' which covers all risks attributable to the enterprise including undiscoverable and unavoidable risks—the defendant will be held liable for all losses attributable to the enterprise. But again, the rational defendant will only take precautions against cost unjustified risks, that is where the cost of liability is greater than the cost of precautions. Where the cost of precautions is greater than the cost of liability the defendant will still be liable but will choose to bear this loss rather than the greater cost of precautions. So at a crude approximation the prevention incentives for both rules are the same.

In two respects, however, strict liability may be more efficient than a fault rule. First, under a fault rule the only prevention incentive on the defendant is to avoid liability by taking reasonable care. Once he is sufficiently careful he will not be liable and his level of activity will be irrelevant. Under a full strict liability regime on the other hand, the defendant is forced not only to consider the cost-effectiveness of his level of care/precautions but also to evaluate the optimum level of activity since, even with the utmost care, he will still incur liability for unpreventable risks and the cost of this liability will be proportionate to the level of activity. Clearly then, where the more efficient solution is a lower level of the risky activity by the defendant, full strict liability is the more efficient rule.[5] Secondly, full strict liability may be a more efficient rule because by internalizing the costs of risks which are unknown or unavoidable given the current state of the art, it provides an incentive to defendants to search for risks or control devices hitherto unknown—the so-called 'technology-forcing' advantage of strict liability over a fault rule.[6]

[4] R. Posner, 'Strict Liability: A Comment' (1973) 2 *J. Legal Studs.* 205, 206–8.

[5] S. Shavell, 'Strict Liability Versus Negligence' (1980) 9 *J. Legal Studs.* 1–4.

[6] M. Katz, 'The Function of Tort Liability in Technology Assessment' (1969) 38

The central concept of the 'cheapest cost avoider' is really an amalgam of abilities to weigh, control, and even insure against a risk. Strict products liability proposals reflect the crude generalization that 'producers' are prima facie the overall cheapest cost avoiders in this sense [while workers' compensation systems select employers—essentially because of the risk perception problems of individual employees[7]]. But in general it is often extremely difficult to predict which party would be selected by this compound concept; for example, the party in the best position to assess a risk may not be in the best position to control it. Even where a risk is unknown and it seems irrelevant from a prevention point of view on which party in the distribution process liability is imposed so long as the loss is efficiently reflected in the price (for example, liability for an unknown adverse drug reaction might be placed on the dispensing chemist), one party may be better able to discover the risk and thereby emerge as the cheapest cost avoider. The cheapest cost avoider may well vary from product to product, in some cases being the producer (for example, pharamaceutical risks), in others the supplier (for example, transfusion hepatitis), and in yet others the consumer (for example, cosmetic allergies).[8] Moreover, it is not clear what relative weight the various factors ought to have in the selection of the cheapest cost avoider nor, just as importantly, the relative importance of the two limbs in the selection, that is, one party may be the cheapest cost avoider with respect to the prevention limb of deterrence but another with respect to efficient reflection of cost in the product price—see below. [It may also be wondered if the concept of a single cheapest cost avoider, to whom the loss is alone internalized, is the most efficient solution in cases due to multiple man-made causes.]

In this second limb of market deterrence economists point out that instead of the relevant cost-benefit calculation being done by fault liability rules or by the defendant in response to strict liability, it may also be accomplished via the price mechanism of the market. One way the tortfeasor copes with the burden of past

Univ. of Cinn. L. R. 587. But see J. Henderson, 'Coping with the Time Barrier in Products Liability' (1981) 69 *Calif. L. R.* 919, 940–2; R. Posner, *Economic Analysis of Law* 2nd ed. (Boston, Little Brown & Co., 1977), 138–9.

[7] G. Calabresi and J. Hirschoff, 'Towards a Test for Strict Liability in Torts' (1972) 81 *Yale L. J.* 1055, 1068.

[8] G. Calabresi and K. Bass, 'Right Approach, Wrong Implications' (1970–1) 38 *Univ. of Chi. L. R.* 74, 83–8.

liability and the potential defendant copes with the perceived cost of future liability is to reflect it in the cost to the public of the relevant activity or product. As the consumer will now be charged a cost closer to the true social cost of the product there may be a marginal shift to a cheaper (because safer) substitute if it exists. In other words, price deterrence is one form of the 'deterrence via lowered activity' mentioned earlier under the prevention limb, but whereas under the prevention limb it is the result of conscious reassessment by the defendant, in price deterrence it is imposed by external market forces. The reduction in use of the product results in reduced losses from it and a more efficient allocation of resources. Predictably, the enhancement of this price deterrence is one of the major arguments used by advocates of a full strict liability rule, because under that rule there is internalization of the full range of costs which are regarded as risks of, and therefore the responsibility of, the enterprise, including costs due to unknowable and unavoidable risks. This degree of internalization is, by definition, impossible under a liability rule which only extends to negligently run risks. In the latter regime, where even in theory prices do not reflect the full social cost, the same level of efficiency produced via price deterrence under strict liability would only be possible if consumers had perfect knowledge, that is, knowledge of the hidden cost of the product involved in risks for which the responsible enterprise will not be held negligent, such as unknowable risks.[9]

Despite the elegance of market deterrence theory, it is well known that tort law does not produce such simple micro-economic deterrent effects in practice. Why is this?

6.4 *Market Imperfections*

(*a*) *In the General Case* Although critics concede that the economic approach to law is useful in highlighting important general considerations, its detail has been attacked on the basis that it is an attempt to fine-tune a system swamped by outside factors. Some market imperfections which undermine the theory are clear, such as the non-legal pressures on victims to settle for less than the true value of their claim; the fact that in our society as presently constituted, certain costs (notably NHS health care

[9] Shavell, 'Strict Liability Versus Negligence', 4.

costs) will be difficult to internalize; that micro-economic price margins may be distorted or overwhelmed by government action such as selective taxation and subsidies; and that other factors besides profit maximization may motivate large organizations. But since there are other distortions in the disease area which are so gross that they render such fine-tuning impossible, it is not necessary to appeal to the imperfections present in the general case to evaluate the theory in the disease area. In the following discussion of the particular imperfections of a disease case, then, these general market distortions will be ignored. For simplicity the analysis also assumes situations in which it is possible to identify the cheapest cost avoider and therefore 'what should be the cost of what'—in fact, as noted earlier, the most difficult yet central question posed by the market deterrence theory. Thus, for example, it is taken that (*a*) 'producers' are the cheapest cost avoiders of diseases attributable to products such as pharmaceuticals, industrial chemicals, etc., and that (*b*) the action of the victim cannot significantly affect the probability or severity of loss—a 'unilateral' situation typical of disease cases because of the victim's inadequate perception of risk or of his particular susceptibility to that risk (see 6.7).

(*b*) *In the Simplified Disease Case* Assume a defendant can tell from his insurance premium what cost is associated with which particular activity or product (an assumption which rarely holds true, see 6.5). Clearly both the prevention and price deterrence effects of tort liability depend on the accuracy with which the loss can be internalized by the defendant (assuming he is the correct party) in such a way that, respectively, (*a*) he is given an incentive to prevent further loss from *that* particular source; and (*b*) it is reflected in the price of *that* activity or product. Yet this may be impossible for a number of reasons. First, it depends on the tortfeasor's *willingness* to internalize the cost to the relevant product. In imperfect markets such as pharmaceuticals, it may be cheaper simply to pass on the full cost in the price of the product than trying to prevent future costs. The individual prevention effect may then be lost while the price deterrent effect may be substantially eroded by the inelasticity of market demand. A related ploy which such defendants might use with comparable results is to spread the loss over a series of products with relatively inelastic demand curves. Both aspects of the theory also depend on the tortfeasor's *ability* to internalize the cost to the relevant enterprise. But in many latent man-made disease contexts the high turnover of products (for example, drugs, industrial chemicals) means that the tortfeasor is no longer engaged in that activity by the time its costs are sheeted home to him via tort liability. The result is that the latter can only have the effect of distorting the cost structures of new enterprises, if there are any. There can be no deterrence in either aspect of an abandoned risky enterprise [although some flow-on prevention effect may be felt in related activities, for example, the losses which would have resulted had liability been found for the thalidomide injuries could not have been internalized to *that* withdrawn product, but pharmaceutical companies could, nevertheless, not ignore future potential liability for failing to screen for foetal damage]. Even where the activity is still being carried out, the prevention aspect of the theory depends on the tortfeasor's ability to avoid the costs of future personal injuries. On the one hand, liability for man-made diseases may seem more 'avoidable' in this sense than for many accidents, because the former is more likely to arise from conscious design choices, for example, in the development, use, and disposal of products like industrial chemicals. Theoretically, then, such decisions should be more sensitive to the standards of liability enunciated by the courts. On the other hand, the discussion in Chapter 4 showed how the scarcity of case-law in the man-made disease context often results in only vague guide-lines to what is considered adequate preventative behaviour. Thus, even if the tortfeasor is willing to take remedial action, he may well have enormous difficulty gauging how much expenditure on health precautions will be cost-effective.

In addition to inadequate guidance from case-law on what is an acceptable level of risk, there are other causes of this difficulty. Firstly, data on the relationship of health standards to risk is usually wholly inadequate—a factor Ashford describes as the most serious of the 'inherent and severe' market imperfections in the area of health hazards.[10] Secondly, decision makers tend to discount future gains, particularly where these are as highly speculative and distant as the reduction in disease from improved

[10] N. Ashford, *Crisis in the Workplace* (Cambridge, MIT Press, 1976), 19.

handling of health hazards. Thirdly, health hazard control is typically more expensive than accident control techniques (see 6.9) and the tortfeasor will often be reluctant to make such outlays—especially in contracting industries—unless the anticipated reduction in risk is to be reflected in insurance premiums. Due to the dysfunction of the premium-setting system in the health field—see below—this relationship of outlay to premium benefit is also speculative. A further complication is that, in practice, many man-made diseases are attributable not so much to conscious design choices as such but ill-informed ones made incrementally through a diffuse company structure. In such cases the concerted expertise necessary to identify and weigh risks may never be brought to bear on the problem. (A striking modern example is the extraordinary—in retrospect—oversight by the pharamaceutical industry of the risk of foetal damage by pregnancy drugs.)

6.5 *The Imperfections of Liability Insurance of Health Hazards*

The most important erosion of the deterrent effect of tort liability is the role played by liability insurance. Virtually all defendants in disease claims would be insured for the relevant risk, and, in the employment context, this will be compulsory. Of course both the prevention and price deterrence effects of tort liability might, in theory, remain unaffected if the insurance market were able to forecast claims and adjust premiums accurately—insurance would simply smooth the loss to the defendant over a number of years while ensuring that funds were available for compensation. There are various reasons why the market is unable to do this.

In two ways insurers might reconstitute the deterrent effect of tort liability, firstly by direct action and secondly by premium rates. In the health hazard area loss minimization by insurers via the direct action of inspecting and advising policy holders is virtually impossible because of the much greater scientific expertise required in this field. Only the crudest level of advice is therefore possible or apparently attempted. Nor would it often be financially worthwhile for an insurer to develop such expertise. Most serious man-made diseases have significant latency periods during which an insurer's competitors would be able significantly to undercut the former's premium rates inflated by the cost of acquiring such expertise. During this time accounts might well be lost and it needs little imagination to conclude that few insurance personnel would want to jeopardize their career prospects by suggesting a highly expensive health hazards advice programme, the benefits of which are speculative and only likely to accrue long after the person's own retirement!

Of course the principal prevention mechanism available to the insurer is the adjustment of premium rates according to the claims experience of the policy holder. Again the theory is simple—the higher the claims experience, the higher the premium and the more likely the policy holder is to take steps to minimize claims through prevention and to reflect the cost of the product in its price. Unfortunately, even in the accident context there are intrinsic limitations to the construction of premium structures which are effective in this way. Atiyah[11] has noted a number of these; most firms are too small for their past experience to be an adequate guide to future claims; the optimal definition of 'experience' itself may be difficult; large claims distort the experience rating inordinately; and variations in liability premiums may be inhibited by the competitive nature of the insurance market especially when masked or restrained by inclusion in a much larger insurance account. The insurer, like the potential defendant, may find the case-law guide-lines to liability—especially in disease cases—patchy and vague.

Even more critical is the problem of optimum insurance categories. To retain the full deterrent effect of liability insurance, premiums should in some way indicate to the policy holder the costs of each activity. This means, for example, that a pharmaceutical company should be able to gauge the liability costs arising from each of its products, say two different tranquillizers, so that it can make relevant adjustments in its prevention efforts with respect to each and reflect the loss of each product in the price of that product. In the real world, however, it is not worthwhile for insurers to draw insurance categories as narrowly as such individual product markets. It is simply too costly. But the larger the category of 'activity' drawn by insurers—for example, 'all tranquillizers marketed by the policy holder'—the less chance there is that individual product liability costs will be correctly ascribed by the policy holder. This is what Calabresi terms

[11] P. Atiyah, 'Accident Prevention and Variable Premium Rates for Work-Connected Accidents' (1975) 4 *Ind. L. J.* 90–6.

'externalization due to insufficient sub-categorization'.[12] Indeed, in practice, insurers usually apply the much cruder 'second-best' solution of classified rating—pooling the experience of a *series* of policy holders, often across an entire industry classification such as 'pharmaceuticals'. This simplifies the insurers' problem of pin-pointing losses to particular activities, although it does not eliminate it. More importantly, it rarely achieves prevention incentives in the policy holder (since action by him will not affect the rating unless he is a dominant enterprise within the classification) and can solely be justified in terms of a quite crude 'second limb' or 'price deterrent' effect—here not between competing products within a defined activity (for example, tranquillizers) but between competing industries (for example, pharmaceuticals and any substitutes).

However, even if premium categories could be drawn sufficiently narrowly the greatest barrier to effective experience (or classified) rating for health hazards is the time-lag problem, which severely restricts how accurate the actual premium level charge can be. Even with accidents, current experience ratings must be set on past experience, usually a period of at least three to five years of past, out of date experience. Often this is an inadequate guide to claims in the current policy year because of changes during the relevant period such as a change in the nature or level of risk of the activity undertaken. But if this is the case for accidents, consider the problems facing risk assessors with respect to man-made diseases. The fundamental and overwhelming problem is one of latency, or, in insurance jargon, of 'long-tail liability' emerging well after the policy year. [A related minor problem of 'long-tail liability', occurring after the initial award of damages, now arises generally in personal injury cases because it is now possible to obtain later awards to cover post-trial deterioration.[13]] Insurance policies typically cover 'injury or disease caused during the period of insurance' and although, as Chapter 2 mentioned, there is a fierce current debate between insurers in the United States as to whether a manifestation or exposure theory should be applied to such words, the exposure theory seems to be generally accepted by

[12] Calabresi, *The Costs of Accidents*, 145; see also 96, 104.

[13] Administration of Justice Act 1982, c. 53, s. 6. On the additional complexities for insurers of health risks see P. Neild, *Byssinosis: The Lancashire Disease* (London, Chart. Insur. Instit., 1982), 45.

insurers in Britain.[14] Thus, if the insurer was on risk during, say, 1950, he will be responsible for claims arising out of exposure in that year even if symptoms did not become manifest until 1980. Of course by that time the policy holder may no longer exist, may have abandoned the relevant line of risky business, or switched insurers. Clearly then, even apart from objections based on good accountancy principles, the insurer cannot rely on future increases in premiums charged to the particular client to cover past losses. A sufficient reserve should ideally have been set aside (that is, raised through premiums) in 1950 to cover losses not arising until 1980. How then does the risk assessor anticipate in 1950 the level of claims in 1980? In the purest form of a strict liability regime, that is one which covers unknown and unknowable risks, the problems of forecasting future liability are horrendous. Thus, in Britain in the few pockets of stricter liability which can be exploited by the disease plaintiff—principally statutory duties with respect to the control of industrial dusts, etc. (see 4.7)—accurate risk evaluation has been impossible and long-tail liability has caused considerable problems for liability insurers.[15] But even under a fault system where the defendant will not be held liable unless the risk was at least known in 1950, the insurer is in no position to assess accurately the level of risk. Moreover, the policy holder may be reluctant to pass on what information he has, indeed in many cases liability is imposed where the defendant's negligence is constituted by a failure adequately to appreciate the risk to which he was exposing people.

This latency problem is exacerbated by the diversity and rapid turnover of the ever-increasing variety of products and processes hazardous to health. It is quite conceivable in a large proportion of disease hazard cases that the hazard no longer exists by the time any claims experience starts so that insurers in the main must rely on the crudest predictions of future claim rates in the setting of health hazard premiums. How, for example, is an insurer to gauge the risks presented by a new drug, the size of the class of potential victims (depending as it does on market success), the likelihood of a finding of fault, and the distribution of future claims over time especially since any adverse reaction may show widely varying

[14] E. Whitmore, *Employers' Liability Insurance* (London, Pitman, 1962), 82, 196; Neild, *Byssinosis: The Lancashire Disease*, 42.

[15] Neild, *Byssinosis*, 42.

latency periods among its victims? Even when claims experience is available at the time of premium setting there are intrinsic limitations to the guidance this can give the health hazard assessor about future rates of successful disease claims. This is because disease claims experience is particularly sensitive to certain extrinsic developments. Of course, a number of influences affect accident and disease claim rates in the same way, such as changes in judicial attitude to a certain class of plaintiff or risk (for example, occupational claims). But there are at least three special ways in which the claims curve for diseases can be dramatically altered. At a particular time a test case may succeed in having a controversial issue of medical causation or fault accepted. Since many disease claims are duplicative this can lead to a flood of comparable claims, a phenomenon generally restricted in the accident field to mass disaster cases such as plane crashes. A second discontinuity in the experience curve for diseases is the surge of claims which may follow the sort of concentrated publicity campaigns recently seen in cases of man-made disease caused by particular pharmaceuticals and industrial dusts. Union campaigns for the scheduling of an occupational disease under the Industrial Injuries Scheme can also lead to similar sharp rises in disease cases. Moreover, unlike accidents, there are a number of factors influencing claims consciousness especially those relating to medical causation and attribution which would be particularly sensitive to publicity (see 5.6). Finally the scheduling of an industrial disease would generally result in a sudden reduction in the number of common law claims unless accompanied by restrictive conditions. In the latter case it might be anticipated that the level of claims could even increase, facilitated by the acceptance of medical causation implicit in the scheduling decision. This is certainly what happened after the severely limited scheduling of occupational deafness (see 4.1). Either way, scheduling can be expected to have a marked impact on claims experience. In contrast, curves of accident tort claims do not show this sort of discontinuity because the parallel compensation systems—tort and industrial injuries—are both 'open', that is, the latter is not confined to scheduled injuries. In short, then, even if the premium setter has at hand past claims experience data—which he could then in theory compare to exposures in earlier years, deciding the appropriate premium for the current years' exposure—the data may in practice be so sensitive to extrinsic events that its use in any predictive way is rendered virtually impossible.

6.6 *Market-Place Results*

Earlier it was noted that one practical result of these enormous risk-forecasting problems for health hazards has been that inadequate cover was taken out by potential defendants and inadequate reserve formation was made during policy years by insurers to cover later liability (see 1.4, 5.6). The indeterminate potential of environmental disasters makes the exhaustion of insurance cover particularly likely in that area. Another practical result has been that in some areas insurers have refused cover—no doubt one of the most forceful forms of economic deterrence. For example, by 1918 some life insurance companies in North America were refusing to insure asbestos workers.[16] The most striking examples in the area of liability insurance arise from the volatility of the legal environment in product liability in the United States during the past two decades. Here the insurability of health risks has become a major worry with many insurers refusing cover especially for particular products such as asbestos. In one case, the refusal of manufacturers' insurers to cover liability arising from a government swine 'flu vaccination scheme led to federal legislation channelling any strict liability to the federal government with a right to recover against the manufacturer only in cases of negligence.[17] At present there is little evidence of insurers withholding cover for health risks in Britain—no doubt because of the comparative rarity of claims and the relatively greater stability of legal liability rules. However, much of the strong and now apparently successful opposition to proposals for pure strict liability for products (including liability for undiscoverable risks) has been based on the argument that it would render certain socially useful development risks—such as those involved with new drugs—uninsurable because of the imponderables such strict liability for unknowable risks involves for premium setting.[18] It follows that the fundamental uncertainties surrounding the width

[16] H. Glasbeek and S. Rowland, 'Are Injuries and Killing at Work Crimes?' (1979) 17 *Osgoode Hall L. J.* 506, 592.

[17] M. Franklin and J. Mais, 'Tort Law and Mass Immunisation Programs: Lessons from the Polio and Flu Episodes' (1977) 65 *Calif. L. R.* 754.

[18] J. Fleming, 'Drug Injury Compensation Plans' (1982) 30 *Amer. J. Comp. Law*, 297, 311.

of the liability set out in the 1985 EEC Directive will generate corresponding uncertainty among insurers about how to forecast future liability in this area and this may well lead to reluctance to offer policies in certain areas.

6.7 *Implication for Deterrence-Based Defence of Tort*

Apart from these market-place results, the characteristic premium-setting difficulties for health risks have important implications for the deterrence debate. First, it would seem that, of all areas, man-made disease illustrates most persuasively the minor role, if any, which tort liability can play (*a*) in specifically motivating those individuals in a position to reduce risks to do so, and (*b*) in achieving price deterrence through the market. Even in cases where liability can be successfully fixed and costs internalized to the appropriate person, the cheapest cost avoider, the deterrent impact on that tortfeasor is lost through a system of liability insurance which is incapable of re-establishing the deterrent effect via effective experience or classified rating. The tortfeasor's incentive to avoid future losses is undermined while price deterrence is severely limited by the difficulties in targeting losses to the correct product especially in areas of high product turnover and long latent diseases. Added to these rate-making difficulties in the real world are the general market imperfections noted at the outset including the difficulty and vagueness of the process of selecting a single 'cheapest cost avoider'. Of course, the greatest imperfection is that the majority of losses will never be internalized in the first place because causal proof is impossible. As Pierce notes, this has 'led to an almost complete breakdown in the tort system as a mechanism of internalizing [man-made disease] costs'[19] and it is presumably for this reason that even supporters of using liability rules to effect economic efficiency note—albeit in passing—that this may not be possible in the area of certain health risks such as pollution disease.[20]

In short, the deterrence value of tort liability fails most dramatically in the disease context: the degree of market deterrence of health hazards via this route is, and can only be, negligible. This, in turn, means that the abandonment of tort here would not

[19] Pierce, 'Encouraging Safety', 1298.

[20] See e.g. F. Michelman, 'Pollution as a Tort: A Non-Accident Perspective on Calabresi's Costs' (1971) 80 *Yale L. J.* 647, 649.

remove any significant safety incentives from the market, even if it is conceded that, in theory, tort could optimize the deterrence goal if it worked effectively. Moreover, in practice the opposite is true: in any system the costs of disease must be allocated. At present most man-made disease costs are focused on victims, their families, and the community in general via systems such as the National Health Scheme and social security benefits. But in the vast majority of cases the victim is either unaware of his susceptibility to the health risk (which cannot be screened for or at least not screened for most cheaply by the victim) or unable to assess adequately the magnitude of the risk, so that, whatever the confusion about who else would be the cheapest cost avoider, it can safely be said that it is certainly not the disease victim on whom the loss at present typically falls. In terms of deterrence theory, therefore, the present system in practice is the worst of all possible worlds in the area of man-made disease. The great bulk of man-made disease costs remain on people who cannot affect the level of risk—they are, therefore, completely externalized leaving disease-producing processes enjoying a considerable subsidy. Moreover, in view of the intrinsic barriers to reforming this externalizing aspect of tort law—such as the need to prove a causal link to an identifiable defendant—these problems will remain impervious to reform. Man-made disease cases would, therefore, seem to provide the paradigm case where tort can be replaced by a superior compensation system with minimal negative effects on deterrence.

6.8 *Deterrence Via Funding of Non-Tort Schemes*

If tort is abandoned and a non-tort compensation system is set up, the design of the funding mechanisms of the scheme will determine the level of internalization of costs and thereby of deterrence. Thus, for example, the present New Zealand scheme, by requiring no separate contribution from say, health care providers or product manufacturers, completely externalizes the costs of injury from these sources which are thereby being subsidized by other contributors. On the other hand, excessive internalization is generated under the New Zealand scheme by sheeting home to employers the costs of all injuries to employees (except non-work, vehicle-caused injuries), not simply those due to work—and a similar result occurs under more direct sick-pay schemes. A certain level of crude internalization is effected in, say, a no-fault road accident scheme by funding it via a petrol tax rather than general revenue. More focused internalization would be achieved in schemes such as the Industrial Injustices Scheme by classified or, if feasible, experience rating rather than the current flat rate. Of course, there would still be the two principal barriers to an efficient rating system in the general case: the economies of drawing categories narrow enough to effect prevention incentives rather than mere price deterrence; and the problem of multiple causation.

Clearly then, the second and more important implication of the analysis in the preceding sections is that, even when tort is abolished, the particular problem of rate-making for health hazards and the vast residue of disease losses which cannot be internalized because of lack of causal proof will continue to undermine attempts to incorporate deterrence via the compensation mechanism. It is true that a centralized compensation system would produce a much wider data base from which ratings could effectively be set than is generated under a fault system. But any system of deterrence necessarily depends on attribution (whether or not associated compensation benefits do) and in only a relatively few cases of man-made diseases would there be sufficient evidence to pin-point the medical cause either generally (for example, 'tranquillizers'—for classified rating) or more specifically (for example, 'product *X*'—for experience rating). Insistence on a deterrent element linked to each payment of compensation for disease, therefore, inevitably results in extremely limited compensation schemes. Of course this link is not essential, the two goals can be independently pursued within the one scheme. The scheme could be funded from general levies or taxation—achieving comprehensive coverage—with special ratings where medical attribution was possible. This type of scheme may well be the most effective way in practice of controlling the costs of personal injuries. A central co-ordinating bureaucracy could use data collected from its compensatory role to improve the causal attribution necessary for the deterrence based variable funding. Moreover, once the tort regime has been abandoned, more sophisticated handling of multiple possible and multiple cumulative causation is feasible. For example, Pierce cites the case of lung cancer.[21] Under a comprehensive system all victims would be compensated. From the data from these claims it may be possible to ascribe, say, 60 per cent to the effects of smoking and 20 per cent to the environmental and occupational hazards of the steel industry. It would then be possible to effect a rough internalization of these costs via variable ratings on these activities while leaving the costs of untraceable illness to the general fund.

Unfortunately, the establishment of medical causation does not eliminate the risk definition problem of what should be the cost of what; that is, who are the cheapest cost avoiders within an activity. The example used in section 5.3 concerning asbestos cancer illustrates the problem in the typical disease case where the victim is helpless to avoid the disease but it is not at all clear which other party should bear the loss: the manufacturer, builder, or architect? Most disease cases are also complicated by issues of multiple cumulative causation, synergism, and recurrence of disability which all pose great difficulties for the selection of the cheapest cost avoider. Of course, in a non-tort system it is not important to select only one party as the cheapest cost avoider. Because averaging over a number of cases is possible, a number of parties may be selected to contribute funds, reflecting the fact that most personal injuries result from a combination of 'causes'. Thus, for example, contributions to cover asbestos illness may not only be extracted asbestos manufacturers but also cigarette manufacturers because of the synergistic effects of tobacco and asbestos. The basis of apportionment between those selected for higher levies, however, would still remain a problem.

The persistence of these risk definition questions into the funding of non-tort schemes can be illustrated by the thorny problem of whether losses associated with abuse of products such as alcohol and tobacco should be seen as costs of the product enterprise, or ascribed to the conduct of the victim and best deterred by leaving the loss on the victim. On the one hand, producers may better appreciate the risks of abuse, particularly with addictive products, and be well placed to take preventative action such as realistic packet warnings, etc. At the very least the health costs of smoking could be reflected in the price of tobacco

[21] Pierce, 'Encouraging Safety' 1324.

achieving some crude price deterrence. On the other hand, some information on the risks of smoking is fairly general among the adult population. Should an attempt be made to encourage healthier life-styles by leaving such losses on the victim? If the latter course is chosen, this raises the dismal spectre of certain victims either being excluded from the 'comprehensive' scheme, or required to contribute a certain increased rate. In either case it is doubtful whether the deterrent effect would be significantly greater than that already present from instincts of self-preservation. Of course, this last consideration suggests the quite general point that the victim of personal injury will rarely be the 'cheapest cost avoider', so that a system which allows the costs of avoidable injury to remain on the victim produces the most extreme externalization and minimal deterrance. Perhaps this explains why Calabresi asserts that it is more efficient to limit cigarette smoking by increasing the price directly (for instance, by taxation or production limits) than to allow the loss to lie on the victim.[22]

On the more general point of attempting to include some deterrence via variable rating, it would seem that given the variety of possible sources of health hazard, the statistical analysis necessary would be staggeringly expensive—and no doubt highly biased to certain 'high profile' risks. In the end, very few disease risks could be adequately or even crudely internalized. The question policy makers would face is, would it be worth the effort? Even if it was in certain cases, there may well be arguments based on equity against distinguishing between enterprises simply on the fortuitousness of causal proof or ease of administration. Such a scheme might not only inhibit scientific research—largely funded by these enterprises at present—but also encourage even more rapid turnover of products so that enterprises could resist internalization via variable ratings on the basis that the risk no longer existed so could not be deterred. In summary, the degree of deterrence of health hazards feasible under non-tort schemes would seem to be greater, but only marginally greater, than under tort. It should not, therefore, assume a dominant role in the decision whether to erect such a scheme nor in its detailed formulation. Whatever course is adopted, most man-made health hazards will continue to be externalized from the responsible

[22] Calabresi, *The Costs of Accidents*, 274.

activities, the latter being subsidized at the expense of the general community.

6.9 *Addendum: Direct Regulation of Health Hazards*

If tort liability is not worth preserving for its deterrent effect on health hazards, and if deterrence via variable contributions to the substitute compensation fund is also severely limited, the obvious question arises as to how society is to control these risks. The controversies about the relative efficiency of regulatory techniques is beyond the scope of this book, but it is worthwhile to note, as a brief addendum to this chapter's discussion, a number of factors in this area which echo themes in the present discussion about compensation. At the outset it must be conceded that no economic theory suggests that the control of health and safety risks ought to be left completely to the mechanism of market and liability rules. Calabresi devotes considerable space to propounding situations in which specific deterrence by public regulation would be more effective. For example, where there can be a collective decision that an activity, such as phosphorous match-making, is not worthwhile at all it should be prohibited outright.[23] Similarly, even if the object is simply to control a risk there may be cases where these are reached more efficiently via specific deterrence. One example is where there is excess externalization due to insufficient categorization of activities under liability/insurance techniques. Another is where the additional deterrent effect of liability or funding rules is only marginal—the case with health risks from tobacco it would seem (see 6.8). Moreover, those who are otherwise critical of direct regulation concede that—because of the greater and intrinsic externalities pervading the market/liability system in the health hazard area—it is in this context that regulation is supported by the strongest arguments.[24] One example of the particular advantages of public regulation of health hazards has already been noted: that its centralized collection of information—

[23] Calabresi and Hirschoff, 'Towards a Test for Strict Liability', 1076. Such action involves inroads into personal liberty—Calabresi, *Costs of Accidents*, 105.

[24] e.g. A. Nichols and R. Zeckhauser, 'Government Comes to the Workplace: An Assessment of OSHA' [1977] *The Public Interest*, 39, 44. See also J. Mendeloff, *Regulating Safety* (Cambridge, MIT Press, 1979), 14, 260; Ashford, *Crisis in the Workplace* 19; Economic Council of Canada, *Reforming Regulation* (1981), 101–2, 107–8; S. Kelman, 'Occupational Safety and Health Administration' in J. Wilson (ed.), *The Politics of Regulation* (New York, Basic Books, 1980), 236, 238, 265.

especially when linked to a compensation system—can lead to improvements in determining medical causation. This, in turn, can result in indirect prevention of disease by the generation of dissemination of health information and direct deterrence by specific regulation.

What then are the features of direct regulation of health hazards which parallel issues in the compensation debate? Firstly, just as historically the compensation debate has focused on palpable traumatic injury, this has also been the main concern in the regulation area, particularly in the regulation of occupational hazards. Regulatory action is often precipitated by a traumatic disaster such as the explosion at Flixborough in Britain or the cave-in at Farmington in the United States. The action itself often reflects a bias towards safety rather than health in legislation and general deployment of resources in areas such as research and standard-setting, advisory functions as well as enforcement. Secondly, regulators have difficulty in setting health standards because of the problem in attributing medical causation (admittedly simplified by not needing to attribute cause in individual cases) and in predicting dose–response relationships in the typical case where data is sparse, imprecise, and often under the control of the parties to be regulated.[25] Of course, even if it were possible to link specific exposure levels to resultant disease levels, the regulator is left with the general regulatory problem of explicitly placing a value on lives and apportioning regulatory efforts in what can be seen as a 'tragic choice'. A third characteristic of health hazards are that they typically involve continuing exposure to a risk. Just as this has implications for assessing common law liability to pay compensation, it also results in regulation characterized by performance standards. This has the advantage of more direct control of the hazard and it has consequently been argued that there is more promise, at least in theory, for efficient regulation of health hazards than safety risks.[26] However, because performance health standards require costly monitoring and depend on efficient and continuing enforcement by highly trained staff, they are

[25] Pierce, 'Encouraging Safety', 1310; Nichols and Zeckhauser, 'Government Comes to the Workplace', 51; Mendeloff, *Regulating Safety* 61, 64; see also Chapter 4.

[26] Nichols and Zeckhauser, 'Government Comes to the Workplace', 51, 42, 161; Kelman, 'Occupational Safety', 236, 238.

intrinsically more difficult and costly to comply with and to enforce. Another special problem in the regulation of health hazards is that because of the multitude of such hazards, it will be impossible in practice to set standards for all but a tiny minority of them. Yet another is that the impact of health measures will not be realized for many years in the case of latent disease and even then may be considerably underestimated because of the problems of causal proof and data collection. Monitoring the efficacy of health hazard regulation like the efficacy of limited disease compensation schemes is, therefore, difficult if not impossible in any accurate sense. A final common theme between the disease compensation and public regulation of health hazards is that in the latter risk definition questions of who is to be regulated still arise.

7

Implications for the Compensation Debate and Beyond

7.1 *Introduction*

The preceding chapters have outlined the conceptual and practical barriers faced by victims of man-made disease in recovering damages in tort and how these difficulties are not satisfactorily resolved by the introduction of stricter liability or non-tort schemes limited on the basis of cause. The conclusion (5.9) was that only a non-tort scheme based on disability rather than cause could overcome these difficulties. The irony is that, while critics of tort law as a means of compensating personal injuries have argued for its replacement by such a scheme on a 'needs' basis, in neglecting the problems of man-made disease victims under tort they have neglected perhaps the strongest arguments for such a reform. Moreover, even if the other major function claimed for tort-law—deterrence—is addressed, it is found that once again man-made disease provides the most damaging illustrations of the weaknesses in practice of such economic defences of the tort remedy (Chapter 6). In short, it is the issues raised by man-made disease which provide the most powerful support for moves to replace tort with a more effective system by which victims of personal injury receive financial assistance.

This chapter looks at how and why, quite apart from the neglect of these arguments justifying reform, the neglect of man-made disease issues has distorted the compensation debate and reform proposals as reformers continue to concentrate on traumatically caused injury. Adequate recognition of the misfortune of disease here would lead to a clearer view of the many policy choices implicit in suggested reforms and would reveal confusion in related social welfare priorities. If there is a real commitment to a needs basis and to the abolition of cause as a criterion for compensation, this throws up for reassessment the currently accepted principles on which reform proposals have been made for the past two decades. Some tentative proposals are made to illustrate this point. None the less, the final conclusion must remain that there are no simple solutions here or short cuts to an assimilation into the reform debate of the problems raised by disease cases. Put shortly: if disease were to be included in the reform debate with the emphasis it deserves, there would have to be a great deal more discussion and analysis to establish what are the fundamental social priorities on which any reform here should be based.

PART I THE COMPENSATION DEBATE

7.2 *Moves Towards Comprehensive Reform*

In the context of social security reform, Beveridge had noted that 'acceptance of [the needs] argument . . . would avoid the anomaly of treating equal needs differently' and that 'a complete solution is to be found only in a completely unified scheme for disability without demarcation by the cause of disability.'[1] As described at the end of Chapter 5, a comparable needs principle has emerged over the past twenty years from dissatisfaction with the way tort law compensates personal injury victims. It is now widely accepted among reformers that in replacing tort an ideal compensation scheme would treat victims of personal injury alike, regardless of the cause of disability. Despite this broad consensus, even reform proposals which espouse the needs basis typically retain a preference towards accident victims. Principal among these are those developed in Australia and New Zealand.

Australian Reform In the past twenty years Australia and New Zealand have been a focus for calls to replace tort with a comprehensive form of compensation. In 1963 the New Zealand Committee on Absolute Liability had noted that 'there is a case for an accident insurance scheme which would cover all persons who are injured in any way.'[2] This and other developments led to the 1967 'Woodhouse NZ' report which, going beyond its terms of reference of occupational accident and disease, recommended a state compensation scheme to replace tort which would cover all

[1] *Social Insurance and Allied Services* (chm. W. Beveridge) (1942) Cmd. 6404, para. 80.

[2] *Report of the Committee on Absolute Liability* (Wellington, NZGP, July 1963), para. 40.

accident victims irrespective of the cause of accident. This was followed in 1969 by a government 'Commentary' explaining certain alternative options and in 1970 by a select committee report, the 'Gair Committee' report which recommended a more limited scheme.[3] However, by the time the scheme came into operation the newly elected Labor government had re-established the Woodhouse principle of comprehensive entitlement covering all victims of all accidents. The responsible administrative agency, the Accident Compensation Corporation (the ACC), recently noted that the scheme is 'firmly based . . . working well . . . inexpensive to fund . . .' and 'a vital and now integral aspect of New Zealand's life-style'.[4] Yet despite earlier forecasts of imminent extension to non-accident disability, thirteen years after the first piece of legislation was passed this extension has still not occurred.

In Australia the Whitlam Labor Government (1972–5) had decided to institute a similar accident scheme and set up a Committee of Inquiry, also under Woodhouse, to advise on how best to achieve this. The Committee was also later asked to advise on the desirability of extending the scheme to all forms of disability including disease and the consequent 'Woodhouse Australia' report recommended a scheme of comprehensive entitlement for *all* forms of physical disability regardless of cause. However, the report noted that the reform could occur in stages—first covering accidents then disease—and a senate committee reported in favour of actually shelving the disease part of the scheme at least until firmer costings were available.[5] In the event, none of the proposed National Compensation Bill survived the fall of the Whitlam Government in 1975. The policy of the present Labor Government is to proceed by co-operation with the

[3] *Personal Injury: A Commentary on the Report of the Royal Commission into Compensation for Personal Injury in New Zealand* (presented to House of Reps. by leave Oct. 1969), hereafter 'Commentary', *Report of the Select Committee on Compensation for Personal Injury in New Zealand* (laid on table of House of Reps. Nov., 1970) chm. G. Gair.

[4] Accident Compensation Corporation, 'Accident Compensation: A Detailed Commentary on the 1982 Act' [1983] NZACR 504–5, 520 dealing with the consolidating Accident Compensation Act 1982 (NZ) (1982, No. 181), hereafter '1982 Act (NZ)'.

[5] See Woodhouse Australia, para. 3; Parliament of the Commonwealth of Australia, *Report from the Senate Standing Committee on Constitutional and Legal Affairs: Clauses of the National Compensation Bill, 1974* (1975, Parl. Paper No. 142), hereafter 'Senate Report', para. 1.23.

states, instituting in stages a uniform no-fault road accident scheme, increased workers' compensation benefits to match those under the road scheme, extension of workers' compensation to 24-hour cover for accidents to earners, and a final extension of accident compensation coverage to non-earners. Non-occupational disease victims no longer feature on the reform agenda at all.

7.3 *The Accident Preference*

It is not necessary to rehearse all the arguments of 'community responsibility' put by these reformers for the introduction of a public system of compensation for personal injuries based on comprehensive entitlement. What is important here is that at every stage of discussion the needs principle of treating like disabilities alike regardless of cause was unanimously accepted as a self-evident element of the reforms called for by community responsibility. There was no concerted attempt to distinguish disability due to accident as a class more or less worthy of attention than disability due to other causes. Nor is it likely that it could have been for, as preceding chapters have shown, of all classes of disabled those suffering from disease are the most inadequately served by the tort and other systems of compensation even though they form the great majority of disabled. Yet at each stage in the formulation of these proposals, a distinction was made in favour of disability caused by accident. The New Zealand scheme, for example, covers personal injury by accident and excludes other sources of disability except (i) occupational diseases, (ii) unusual adverse drug reactions which fortuitously qualify under the medical misadventure provisions, and (iii) deafness inexplicably included as 'personal injury by accident'.[6] As a consequence, victims of man-made hazards such as environmental pollutants and non-medicinal products such as food, cosmetics, and other chemicals go uncompensated under the scheme. They are relegated to what is usually the illusory remedy provided by tort. Those who cannot establish the man-made origin of their disability are left to the far more limited support provided by social security and sick-pay schemes. Retaining the accident preference in this way has not only the practical disadvantages of demarcation difficulties and costs but it also necessitates the retention of a liability insurance

[6] 1982 Act (NZ), ss. 2, 28, 29. Woodhouse NZ, paras. 17, 290(d) Woodhouse Australia, paras. 353(c), 525.

network to cover those few disease cases which reach and succeed in the courts or settlement processes. The preference also produces distasteful anomalies highlighted in, say, the treatment of widows, which depends on the 'fortuitousness' of the husband's death being due to an accident.

What is the reason for the accident preference? Pressure from vested interests to retain disease within the liability/insurance field cannot be the answer. The income from disease litigation is insignificant to the legal profession. Moreover, as described in Chapter 6, insurers find health cover particularly difficult. Indeed, in submissions to the Woodhouse Committee, the Australian insurance industry actually argued for the step by step proposal to be reversed with the detached sickness scheme being implemented first and (unlike the insurers' proposal for the injury scheme) handed over to the government as insurer on the ground that the resources necessary for the sickness scheme were far beyond the capacity of the insurance industry.[7] Ironically, in rejecting this idea Woodhouse noted that it produced parallel systems differentiated on the basis of cause, a result to be rejected on principle. Of course this reasoning simply emphasizes the greater anomaly in the 'accident scheme first' option which would not have encompassed what Woodhouse himself saw as the 'major field of compensation', disability caused otherwise than by trauma.

Reformers like Woodhouse have generally given no explicit detailed rationale for giving priority to traumatic injuries except the pragmatic one of cost.[8] An accident scheme would be much cheaper than a sickness scheme—Woodhouse Australia estimated it at about one quarter and since statistics on disease were sparse the cost of the sickness scheme may have been even higher.[9] Cost is a rational criterion by which reform priorities can be set. It

[7] Woodhouse Australia, paras. 15, 220, 224, 227–8, 297. On some problems of providing disability insurance see T. Ison, 'Human Disability and Personal Income', in *Studies in Canadian Tort Law*, ed. L. Klar (Butterworths, Toronto, 1977), 425; Atiyah *ACL* 323–8.

[8] Woodhouse NZ, paras. 17, 290; Senate Report, paras. 1.23, 4.1–4.4; *Report of the Royal Commission of Inquiry: Social Security in New Zealand* (Wellington, NZGP, 1972) (hereafter *Social Security in NZ*), ch. 18, para. 98; E. Bernzweig, *By Accident Not Design* (New York, Praeger, 1980), 165. See criticism by T. Ison, *Accident Compensation* (London, Croom Helm, 1980), 29–31.

[9] Woodhouse Australia, para. 483. Other relevant findings on cost are: Pearson, para. 1711; G. Palmer, 'Compensation for Personal Injury: A Requiem for the Common Law in New Zealand' (1973) 21 *Amer. J. Comp. Law*, 1, 11.

makes sense that if all cannot be helped, those who can, should be. Moreover, this applies even if there is no prospect of a step by step extension to comprehensive coverage of all disability. The resultant entrenched anomalies are simply the price to be paid for doing what is feasible in alleviating hardship in a limited area. But the cost consideration does not necessarily lead to this particular form of compromise. The fact of limited resources does not explain why the preferred first and perhaps only step of reform is to provide generous benefits to a class of disabled selected on the basis of cause of disability. Arguably it might make better sense to start with low uniform benefits for all disabled increasing them across the board if and when resources allow. Alternatively, initial preference might be given to the long-term seriously disabled. Cost alone, therefore, is not a satisfactory justification for limiting the state compensation scheme to accidents.

An interesting suggestion has been made by Calabresi. He noted that if compensation were the primary goal there would be no justification in limiting schemes to accidents. He suggested that the reason why compensation for illness is less developed 'even in highly welfaristic countries' is because there seem to be ways to reduce the number and severity of accidents. In other words, he ascribes the preference to the fact that accidents can be deterred while 'usually . . . we apparently do not believe that a different allocation of most disease costs would do much to spur reduction of their number or severity'.[10] Even if the element of bootstrap logic is ignored, the argument does not explain the emergence of the accident preference in the Australasian proposals. For while it is true that Woodhouse asserted that prevention was the most important goal, he rejected techniques for linking compensation to economic incentives for the prevention of personal injuries.

7.4 *The Accident Preference as a Legacy of Effective Tort Liability*

It seems that the accident preference is probably derived from the fact that in the past academic lawyers and other reformers who have arrived at the idea of a 'needs' basis for compensation reform have generally done so after an examination of tort as a compensation mechanism. Because tort liability is effectively

[10] G. Calabresi, *The Costs of Accidents* (New Haven, Yale University Press, 1970), 43–4 and footnote 10 therein.

confined to cases of traumatic injury the focus on tort produced a focus on accidents. Critics of tort were, therefore, misled by its poor performance in compensating accident victims into seeing a particularly 'urgent need'[11] to replace it with a public system of *accident* compensation while overlooking the fact that the absence of case-law on man-made disease suggests an even greater need for reform in this area. The critical evaluation of tort was not taken far enough. Even if reform is primarily aimed not at the needs basis which is professed but at the substitution of a no-fault system in cases to which tort formally applied, then reform must also cover man-made diseases. In maintaining an accident preference at this stage reformers seem to have been guilty of one of their own major criticisms of the supporters of tort law: that in concentrating on those who get to court one ignores the majority who do not get full or any damages from the system because of its informal barriers and vagaries of operation. Critics have been eloquent in describing how the majority of accident victims are excluded in this way yet they have given scant attention to the greater number of victims of man-made disease who fail to obtain tort damages. By ignoring man-made disease as a class, personal injuries become roughly divisible into accidents for which tort proved inadequate and should be replaced and non-traumatic disability arising from 'natural' causes which never gave rise to the possibility of tort claims[12] and which could, therefore, be ascribed secondary priority in the pragmatic implementation of a compensation plan designed ultimately to reflect a needs basis.

The most important result of this unrigorous thinking and its resultant tort-derived accident preference is that it raises to an extraordinary degree the baseline of benefits considered adequate under any new *'replacement compensation'* scheme. In general,

[11] Woodhouse NZ, paras. 17, 290. Compare the view of *Social Security in NZ*, that extension of earnings-related benefits to sickness was 'not urgent', ch. 18, para. 98.

[12] The approach is implicit in Woodhouse NZ, paras. 17, 290; Commentary, para. 110; Gair Committee, paras. 8, 18; Woodhouse Australia, paras. 3, 58, 220 355(g) (although some awareness of man-made disease can be glimpsed in Senate Report at paras. 10.1, 10.23); R. Gaskins, 'Tort Reform in the Welfare State: The New Zealand Accident Compensation Act' (1980) 18 *Osgoode Hall L. J.* 238, 264, see also 259; G. Palmer, 'Compensation for Personal Injury: A Requiem for the Common Law in New Zealand', 1, 44; G. Palmer, 'Accident Compensation in New Zealand: The First Two Years' (1977) 25 *Amer. J. Comp. Law*, 1, 25; Atiyah, *ACL* 4, 364, 443, 498, 507–8, although see 499.

reformers seek to provide what I will loosely call 'standard of living' benefits to approximate as near as is feasible under an administrative scheme to the compensatory position in tort damages. The hallmarks of this approach are the provisions of earnings-related benefits, payment throughout the full period of incapacity, and often some benefits for non-economic losses.[13] The origin of this compensatory notion is generally clear. Historically, the law did not intervene in the case of 'natural' misfortunes such as sickness, unemployment, etc. Here retaliation by the victim was unlikely. In the area of accidental injury public order may have been threatened because the victim could point to an individual 'wrongdoer'. In such cases concepts developed to allow the victim a remedy in monetary damages. In this 'corrective justice' or 'compensatory' context it seemed quite sensible to require sufficient benefits to restore the victim's pre-injury standard of living in so far as money could. Similarly, inasmuch as no-fault industrial injury schemes were initially seen as a substitute for the inadequacy of this tort remedy in practice in the industrial setting, where there is incorporation of above-subsistence benefits (earnings-related benefits, non-economic loss, etc.) these can be seen as an attenuated application of this compensatory common law principle. The rationale of reformers is that it is only fair—or at least politically feasible—to replace the existing regime if the new benefits approximate as closely as possible to the expectations generated by the tort remedy (and by industrial compensation). This is argued even though it is admitted that only a small minority of accident victims receive any tort damages under the present regime (Pearson's estimate is 6.5 per cent). Ironically, then, the uncritical carry-over into the replacement scheme of the very term 'compensation' serves both to mask and to facilitate this important extension of preferential benefits.

To reiterate the reasoning leading to the pronounced accident preference: initially reformers concentrate on tort and this produces a concentration on accidents. Tort is found inadequate as

[13] See e.g. Woodhouse NZ, para. 4; Gair Committee, paras. 14, 77; Woodhouse Australia, paras. 9, 257, 343–4, 365; Pearson, paras. 800, 1012. For North American acceptance of a 'standard of living' approach see also P. Weiler, *Reshaping Workers' Compensation for Ontario* (report submitted to R. Elgie, Minister of Labour, Nov. 1980), 47; Ison, 'Human Disability and Personal Income', 429; R. Keeton, 'Compensation for Medical Accidents' (1973), 121 *U. Penn. L. R.* 590, 609.

a compensation mechanism. To justify abolition rather than reform of tort, a no-fault principle and 'needs' criterion of reform is developed. Yet this is not rigorously applied when it comes to the design of reform proposals. Instead, the tenacious hold of traditional preoccupations reasserts itself producing an accident preference in reform proposals. Reform itself is limited—accidents are to be dealt with first. Since the overwhelming number of beneficiaries of the replaced regime were in practice accident victims, this limitation of the new scheme's coverage to those disabled in this way can seem less exceptional. Moreover, when the level of benefits is set, they are set to reflect the idea of 'just' compensation developed under tort law (and its substitute, industrial compensation). The perceived necessity of protecting the interests of those accident victims with a remedy under the replaced regime provided a convenient justification for the provision of generous 'standard of living' benefits to the large class of accident victims who would not have had such a remedy. This can then appear both a reasonable and generous rationalization of existing resources; even though it in fact not only formalizes the preference to accidents which had informally operated under tort liability as it operated in practice, but extends that preference to a larger class for which there is no independent reason to accord *compensatory* benefits as opposed to straightforward financial assistance.

7.5 *Including Disease Throws Doubt on Differential Treatment of Personal Injuries and Other Misfortunes*

If an examination of the accident preference and its origin reveals that accident compensation schemes provide generous 'standard of living' benefits because reformers see the reform here primarily in terms of a limited rationalization of existing compensation schemes (systems which in practice focus almost exclusively on the misfortune of accidental personal injury) and not in terms of a broad co-ordinated social policy of 'needs' as professed, then what would be the results if broader principles were to be firmly applied? What would the effect be, for example, if the 'needs' basis of reform were to be given the centrality that is often claimed for it?

Earlier it was argued that it is typically the victim of man-made disease who is worst served by the regime of tort liability in practice. Even if the focus of reform is limited to the failure of tort, coverage of man-made disease in any replacement scheme seems essential. What is more, as described earlier (5.9), the distinction between man-made disease and sickness from other causes is not administratively feasible so that any scheme that seeks to cover the former must, by practical necessity, cover the latter, regardless of whether the distinction is itself meaningful in scientific terms or desirable in policy terms. To cover man-made disease, then, one must abandon the natural/non-natural distinction between injury misfortunes. If the principle of reform is the broader social ideal of a needs basis, disease should be covered and there is not even a need to appeal to administrative feasibility to arrive at this conclusion. If need is the basis then the origin of the injury, accident or disease, is irrelevant.

If there is a commitment to a needs basis and to the inclusion of disease in any compensation scheme this has profound effects—it leads eventually to a questioning of the current artificial or at least unexplained differential in policy between disablement and other misfortunes. How is this so? It is clear that in a comprehensive scheme which covers not only accidents but disease the vast majority of beneficiaries will not have been eligible for the generous benefits under the replaced tort/industrial compensation regime. Approximately 90 per cent of cases of incapacity are due to non-traumatic causes (see Figure 7.1), yet, as previous chapters have shown, recovery of tort or industrial compensation benefits by those disabled in this way is relatively rare. The consequence of including disease in a scheme, therefore, is that it will be much harder to justify treating disablement preferentially to other misfortunes on the basis that some beneficiaries had formally been entitled to generous compensatory 'standard of living' benefits. In a comprehensive scheme this group would be tiny (see Figure 7.1). Arguments based on the protection of their rights and the characterization of the reform as a 'rationalization' of existing compensation schemes lose much, if not all, of their force. It would become more arguable that the concerns of the very few are outweighed by the public benefit that abolition of their rights would produce. Once protection of these rights is abandoned as an integral aspect of reform, what should dictate the level of benefits which is 'just' for the disabled? (The term 'compensation' is no longer appropriate.) An explanation as to why, if at all, the

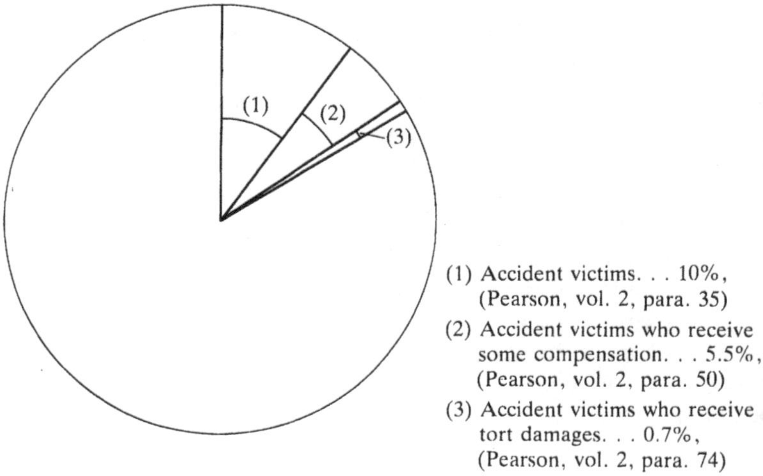

Fig. 7.1. Contributions to Disablement

misfortune of physical disablement should be treated in terms of benefit levels preferentially compared with other misfortunes is called for. Of course, such a system may be funded on an insurance basis with higher benefits going to those who have made higher contributions. But this simply reformulates the question: why should society be so solicitous of the plight of the disabled that it constructs a compulsory insurance scheme to protect the individual's earlier standard of living while not providing the same type of insurance protection to, say, the victims of involuntary unemployment? Why should a victim of disablement receive earnings-related benefits from a state scheme and the victim of unemployment or a dependent wife who has been deserted get far more modest support? What is special about the misfortune of physical disablement? This is the broadest contribution to the reform debate which a focus on, and commitment to, disease compensation makes: it exposes the need for an evaluation of social welfare priorities across a much broader spectrum of misfortune than simply personal injuries and disablement (see below). It also highlights how the continuance of the limited accident focus of reform distorts and disguises these issues to the detriment of rational policy making.

7.6 *Ramifications of the Accident Preference Within the Compensation Debate*

Assume for the moment that some rationale can be found to support a separate public system of preferential benefits for the disabled and that the needs basis is genuinely adopted so that no distinction made on the grounds of how the disability arose is considered justified in principle. From this starting-point it is easier to see that the distortions within, what I shall still loosely call, the 'compensation' debate are caused by the current focus on accidents and its resultant tort-derived idea of what the benchmark of just benefits should be in any reform proposals.

(*a*) *Preclusion of Ultimate Comprehensive Goal* The most obvious problem with a tort-derived concept of what is a 'just' level of benefit is that its costliness may preclude the ultimate comprehensive goal. The reform may be halted at the accident stage—as seems to have been the case in New Zealand. Indeed, the current vogue elsewhere for step by step reform—typically with a first step of a no-fault road accident scheme like that suggested by the Pearson Commission or the New South Wales Law Reform Commission—may in fact be halted after the first step and not even reach all accidents let alone disease. This is because, apart from work injury (already served by a no-fault scheme), road accidents are the major area where tort liability was sufficiently real in practice to precipitate compulsory insurance. Ironically, the pattern of effective tort liability tends to guide and limit the extent of feasible reform because these pools of insurance money may readily and politically painlessly be converted to fund a state-run no-fault road scheme.[14] But once these are used to finance the first step of reform, there may be a lack of the political will necessary to raise or divert from other public programmes the additional funds necessary to cover comparable benefits for the remaining victims of accident and for disease victims, even though these form the vast majority of the disabled. In other words, there may be a fundamental policy choice unwittingly being made here. In 'selling' the idea of a no-fault system of benefits by the almost magical creation out of

[14] Palmer describes how exploitation and rationalization of existing funds was a basic Woodhouse strategy, 'A Requiem for the Common Law', 25, 30. Gaskins, 'Tort Reform in the Welfare State', 271, notes how the focus on existing funding arrangements inhibited extension to non-auto injuries to non-earners in NZ.

existing funds of such generous benefits for the 75 per cent of road accident victims who would have received nothing under the former tort regime,[15] reformers may, in fact, be choosing the goal of rationalizing at a generous level existing compensation funds within a limited area at the price of abandoning the espoused ultimate goal of comprehensive cover of all disability. Of course, as disease is typically left to the last stage of step by step strategies, it is likely to be the first area precluded by a reform truncated by lack of funds. It may be that there *is* an implicit pragmatic preference for short-term rationalization of compensation funds within a limited area of disability. There may even be a different attitude to the compensation of traumatic injury itself[16] or to injuries of a particular class such as pharmaceutical injuries—despite the vigour with which reformers reject such causal distinctions. But in either case these policies ought to be spelt out if only for intellectual clarity and to dispel false hopes of ultimate comprehensive cover.

The goals of short-term rationalization at generous benefit levels and ultimate comprehensive cover may not be mutually exclusive. It has been argued that the more generous benefits in the limited first step—such as those for accidents in New Zealand—can serve as positive pace-setters for other areas, even for social welfare benefits in general. This is the beneficial 'legacy of the common law' forecast by Palmer in New Zealand.[17] The problem is that without an explicit and well-thought-out basis, this flow-on effect may occur in areas where it was unintended: for example, in New Zealand areas affected in this way include other income maintenance areas such as pensions, and benefits such as home alterations.[18] Alternatively, as noted earlier, financial constraints may block the flow-on to intended areas, creating unintended entrenched anomalies. In either case the potential for the distortion of social

[15] Pearson, Table 5, p. 24; para. 994.

[16] See e.g. D. Elliott and H. Street, *Road Accidents* (London, Penguin, 1968), 250: 'it is a fact that the concern of society about the traffic victim is of a different order from its anxiety about the cancer victim', although arguably this may be attributable to existing patterns of *effective* tort liability and is not necessarily a suitable criterion for a post-tort regime (see 5.6).

[17] 'A Requiem for the Common Law', 3. See generally Palmer, *Compensation for Incapacity*, ch. 19, and Gaskins, 'Tort Reform in the Welfare State'.

[18] A. Ogus, P. Corfield, and D. Harris, 'Pearson: Principled Reform or Political Compromise?' (1978) 7 *Indust. L. J.* 143, 149; Ison, *Accident Compensation*, 22, respectively.

welfare policy by an ill-thought-out 'pace-setter' strategy is high.

If, on the other hand, there is an intrinsic and inescapable trade-off between short-term rationalization of compensation funds within a limited area of disability and the comprehensive goal, an interesting possibility arises which turns conventional reform wisdom on its head. This is that a firm commitment to the comprehensive goal may require any step by step strategy for reform to employ *a first step which covers disease rather than accidents*. As already noted, it is in the accident area that organized vested interests are strongest. In many jurisdictions, for example, workers' compensation provides earnings-related compensation for at least the first twenty-six weeks. Since approximately 95 per cent of accidental injuries recover within this period,[19] reform proposals which provide benefits at a lower level have been vigorously resisted by organized labour in those jurisdictions. Abandonment of the earnings-related principle in a first step accident scheme would, therefore, hardly seem viable politically. Strangely then, the comprehensive goal may best be served by a step by step strategy installing benefits at less generous levels (say along the lines of the British Industrial Injuries Scheme) beginning with *non-traumatic* injuries. Since the vast majority of this class has no effective claim under tort or industrial compensation, the 'abolition of rights' argument would be defused while the associated lack of vested interests would assist the reform. Moreover, being the most costly, a first stage covering disease may be a better test of basic political commitment to the comprehensive goal. In the longer term, the anomalous preferential treatment of disease victims might then generate pressure for extension to accidents and undermine the arguments of the privileged section of the accident class (tort and industrial compensation beneficiaries).

(*b*) *Trade-off Between Wealth Distribution and Deterrence Goals*

Another problem with the tort-derived idea that benefits ought to protect an individual's standard of living is that the administrative practicalities of funding may require a choice to be made between policies of wealth distribution and deterrence. Unlike flat rate benefits, a system of 'standard of living' benefits does not

[19] See e.g. Woodhouse Australia, para. 399.

preferentially assist the poor, and, in fact, unless funding of a scheme is carefully structured, redistribution in the opposite direction will occur. If no distribution in either direction is desired then a beneficiary with a previously high standard of living should have been contributing proportionately more to the fund. This could be done by variable direct contributions. It might also arguably be achieved in a crude way if funding was drawn from income tax revenue.[20] If, however, funding only comes from direct levies on activities seen to contribute to personal injuries in an attempt to channel and optimize the deterrent effects of the scheme (see 6.2–6.8), then redistributive effects towards the higher earner will be considerable.

This problem arises even in limited schemes. For example, a road scheme funded by a flat petrol tax but providing standard of living benefits will be regressive, that is, the poor will draw proportionately less from the scheme for what they contribute. Similarly, variable funding by industry for disability caused by a range of industrial (or product-related) risks will also be regressive under such a system of benefits in much the same way, as the tort/liability insurance system operates regressively. Theoretically it seems possible to combine variable rates from injurers (more correctly 'best cost avoiders') reflecting deterrence goals and variable rates from beneficiaries of the standard of living benefits so that he 'pays his way'. In practice this may be administratively impossible, producing a funding dilemma which can only be resolved by compromising or even by completely abandoning either the deterrence or insurance principle. The most popular solution adopted by reformers who adhere to the standard of living benefit regime seems to be to abandon the insurance principle with consequent regressive effects. Alarmingly, this often goes unremarked—for example, the earlier Australasian reformers displayed a surprising confusion on the point, repeatedly referring to the goal of social *insurance* (with earnings-related benefits) yet approving funding mechanisms which did not reflect this goal at all.[21] Nor did the abandonment of the insurance idea seem to be attributable to

[20] Palmer, *Compensation for Incapacity*, 327, but cf. Atiyah, *ACL* 199. Conversely, a flat rate benefit from general taxation redistributes to the poor which some may find problematic, e.g. J. Keeler, 'Report of the National Committee of Inquiry into Compensation and Rehabilitation in Australia' (1973–6) 5 *Adel. L. R.* 121, 129.

[21] Woodhouse NZ, paras. 4, 243, 279, 290, 314; Gair Committee, paras. 78, 110; Woodhouse Australia, paras. 243, 344, 493.

any well-thought-out implicit rationale in favour of the deterrence goal because, typically, of the machinery available to optimize deterrence via funding that which was selected was exceedingly crude.

(*c*) *The Problem of Non-Economic Losses* A third difficulty with a tort-derived notion of what level of benefits is just is that it encompasses payment for non-economic losses. Were a benefits system simply designed to replace an earner's loss of earnings, assessment problems—such as what to do about victims who were self-employed or whose earnings were likely to have increased or decreased with age—would be relatively manageable. But since such a scheme would exclude completely those who are non-earners (a majority of the population[22]), and since such a scheme does not compensate earners for partial disabilities which have no impact on earning capacity (such as deafness in one ear or reproductive injury), it is thought correct and in keeping with tort notions of compensation to incorporate benefits for non-economic losses.[23] The problem with this aim, however, is that it is not administratively feasible to provide the sort of individual assessment attempted under tort, nor has there been developed a workable formula which compensates non-economic losses in an equitable manner (see 7.7). More broadly, the provision of benefits for non-economic losses produces particularly distasteful anomalies in limited schemes for personal injuries such as that in New Zealand where, for example, a person may receive compensation for the disappointment felt by a disability barring a previous hobby while a non-occupational cancer victim receives nothing at all from the scheme. Even in a comprehensive scheme demarcation problems are raised: if the fund is to cover non-economic losses due to the physical condition *per se*, does this mean that, say, the infertile would quality for benefit? (See 7.7.) Benefits for the non-economic losses associated with physical condition may also seem

[22] Where the unemployment rate is 5.6%, 57% of the population are non-earners: New South Wales Law Reform Commission: *A Transport Accident Scheme for NSW: Accident Compensation Working Paper I* (May 1983), para. 7.2.

[23] Despite apparent distaste (Keeler, 'Report of the National Committee of Inquiry', 125; Palmer, *Compensation for Incapacity*, 221), Woodhouse accepted that there should be benefits for non-economic loss: Woodhouse NZ, paras. 60, 198; Woodhouse Australia, paras. 365, 373. Compare minority opinion in Pearson, paras. 457–64.

anomalous when it is remembered that personal injury and disease victims would already be preferentially treated by generous benefits for economic loss. Why provide benefits for the pain and suffering of deafness but not the real economic loss, let alone pain and suffering, of unemployment? Again the point can be made that the inclusion of disease within reform proposals would drastically weaken the appeal of tort as a guide to what a 'just' level of benefits was. In this specific context it would challenge the provision of non-economic losses in a public system of benefits for misfortune. Why should the system provide for non-economic losses when the vast majority of disabled would not have had a claim for such benefits under existing regimes?

PART II WHAT CAN AND SHOULD BE DONE FOR THE DISABLED?

7.7 *The Wider Range of Options for Disability Reform*

In the preceding section it was assumed that a rationale could be found for having separate and preferential treatment for the misfortune of physical disability even though, on principle, no distinction should be made within that class based on cause. It was clear that the current adoption of a tort-derived notion of what should be the nature and level of preferential benefits in such a scheme leads to significant problems both in a practical sense but also in achieving long-term goals. But earlier (7.5) it had been shown that a real commitment to a needs basis, to the abandonment of causal criteria, and to the consequent inclusion of all cases of disability within the preferential personal injuries scheme severely undermines, if not completely demolishes, the argument for using tort as the guide-line for the type and nature of the benefits to be offered by the scheme. This is because, to the vast majority of beneficiaries, the scheme is not a *replacement* scheme for benefits to which they formally would have had access. Once this tort-derived concept that the new system ought to aim to provide standard of living benefits is abandoned, reformers must face up to the problem of whether physical disability is to receive preferential treatment over other misfortunes at all. For the sake of analysis one might at this stage assume that this threshold question can convincingly be answered, yes: that for some specifiable and coherent reasons society should treat disability differently from other misfortunes. It is, however, not a point easily glided over even for the sake of debate, because without tort as a guide to the range of losses worthy of benefits (e.g. actual wage loss, lost earning capacity, pain and suffering, etc.) and the level of benefits which is desirable (i.e. 'full' 'compensation'), how would one translate the commitment to a preferential treatment of physical disablement into a concrete scheme of benefits? What aspects of disability as a misfortune set it apart from other misfortunes and, in so doing, provide a focus for this preferential system of benefits? Without tort to guide the reform, the compensation debate is opened up to a surprising range of options, including options which are currently dismissed as limited and inadequate in terms of a reform seen, whatever else it is seen as, as a *replacement* for the 'full compensation' provided by tort. Let us look first at the range of some of these options.

Option I: Income as the Focus (see Table 7.1) It may, for example, be decided that the focal misfortune which the state should seek to alleviate here is interruption of the actual or potential income of the victim. In other words, the central issue is seen as the impact the physical injury has had on the victim in his capacity as an actual or potential member of the paid work-force. Certainly this is the loss on which workers' compensation systems originally concentrated, despite the fact that they were seen as replacements for the tort remedy (although arguably this might be explained by the fact that at the time these systems were set up, tort damages did not have the dominant component of non-

Table 7.1 Relationship of Benefit to Earnings and Disablement

Option I: Income as the Focus
(*a*) Flat rate:
Benefit @ £100 – post-disablement earnings up to £100
(*b*) Earnings-related (here, the actual lost earnings approach):
Benefit @ pre-disablement earnings – post-disablement earnings

Option II: Disability as the Focus
Benefit @ £100 × % disablement

Option III: Combined Focus: Disability Benefit Geared to Earnings
Benefit @ % disablement × pre-disablement earnings

economic losses that they do today.)[24] In support of an income-focused system one could argue that, while society restricts funds available for social welfare programmes to modest levels, these resources should be concentrated on economic losses, it being more important or feasible for society to spend money to raise the economic safety net than to alleviate intangible losses such as pain and suffering. But even if the selection of economic loss as the policy focus is settled, there are a number of vital subsidiary issues to be worked out. Should we, for example, be concerned with partial loss of earnings/earning capacity or would it be better to concentrate effort on cases of a total loss—as is currently done, for example, in sickness benefits under social security where total incapacity for work is a prerequisite to entitlement? Similarly, a fundamental and more ideologically fraught question is whether benefits should be at a flat rate aimed at income support or earnings-related aimed at income continuity.

In the past, in both social welfare systems covering a broad range of misfortunes and in systems of industrial compensation for personal injury in particular, emphasis initially centred on flat rate benefits designed to provide a basic floor standard of living and this idea has matured into broad proposals for a guaranteed minimum income.[25] Unless the flat rate (from which post-disablement earnings are subtracted—see Table 7.1) is fixed at a high level (which would set correspondingly high disincentives to rehabilitation), the practical effect of such a flat rate system is that for the majority of claimants benefits will only be triggered in cases of total incapacity for work. One of the major advantages, therefore, of flat rate benefits for physical disability is that they are more easily accommodated within existing welfare benefits which focus on economic loss only, are flat rate and, in the case of non-occupational injuries, focus on total incapacity for work. Another advantage is that under a flat rate system the problems presented

[24] A. Larson, 'The Wage Loss Principle in Workers' Compensation' (1980) 6 *William Mitchell L. Rev.* 501. More than half total tort damages are for non-economic losses, Pearson, para. 81.

[25] Ison, 'Human Disability and Personal Income', 427. For modern support for flat rates see R. Abel, 'A Critique of American Tort Law' (1981) 8 *Brit. J. of Law and Soc* 199, 210–11. On the idea of guaranteed minimum income: Palmer, *Compensation for Incapacity*, 48, 169–70, 325; R. Henderson, 'Criteria for Welfare: Needs or Earnings', in A. Graycar (ed.), *Perspectives in Australian Social Policy* (Sydney, Macmillan, 1978), 177.

by non-earners are much more manageable than under a scheme which seeks to include some sort of earnings-related element.

Looking at short-term goals, the flat rate idea is attractive because it reflects the idea of people having basic minimum needs which, if any, are those which society should seek to protect first—before, say, the financial expectations generated from a victim's pre-disability position in the socio-economic pyramid. In terms of long-term goals, the flat rate income support approach has the advantage of being consistent with two widely differing views. It is consistent with the ideological view that society, at the very least, should never engage in perpetuating economic differentials. It is also consistent with the contrasting view that society *should* in the long term aim at providing mechanisms (for instance, state-organized insurance) by which people can protect their standard of living, because the flat rate system is a feasible first step which can, until the ultimate goal is reached, be augmented by private systems of savings or insurance to buffer the fall in standard of living.

The second income approach is to aim from the start for some earnings relation in benefits. In an earnings-related scheme the idea is to compensate for the loss of actual earnings (reflecting current standard of living) or loss of earning capacity and its associated potential standard of living. Such a system has the advantage that cases of partial loss of earnings/earning capacity can *in principle* be more smoothly incorporated than in a flat rate system where discontinuities are inherent in the system. In practice, there *are* major problems. Although the focus of actual lost earnings might be chosen—in which case non-earners are intentionally excluded—most commentators agree that the focus ought ideally to be loss of earning capacity.[26] But this leads to the most intractable problem in an earnings-related scheme: how to assess cases of partial loss of earning capacity, a question considerably more important in a scheme covering all types of disability because such cases would be numerically much more

[26] Larson, 'The Wage Loss Principle', 525. Actual loss of earnings was initially the formula adopted for the NZ scheme but it was never used—see Palmer, 'Accident Compensation in New Zealand: The First Two Years', 25. The current NZ formula is based on Option III (see below). The bias against non-earners remains: entitlement for them is limited to lump sum non-economic losses; no attempt is made to take account of lost future earning capacity or value of unpaid work, see. e.g. *Re Penn: Decision 732* (1982) NZACR 250.

significant in the long term than in an accident scheme (where they total only 2 per cent of beneficiaries[27]). To assess loss of earning capacity without a case by case evaluation, which is administratively not feasible, some approximation technique is necessary. One is, in fact, to use actual loss of earnings to give a crude reflection of lost earning capacity (illustrated in Table 7.1). Another is to use a formula involving a percentage disablement factor (without an element based on pre-disablement earnings—see Option II). Both have considerable drawbacks: the first, by depending on proof of actual post-disablement income loss inhibits rehabilitation while disadvantaging non-earners and those who had unrepresentatively low pre-disability earnings—such as victims of degenerative disease where attainment of a compensatable level of disablement was preceded by a long period of erosion of earning capacity; the second is at the very least intrinsically biased in favour of white-collar workers who tend to suffer less loss of earning capacity for a particular injury than a blue-collar worker.

A combination of flat rates and some earnings-related benefit is possible: although state benefits were typically set up at a flat rate, at a later stage and for complex reasons, limited (that is, partial) earnings-related additions were made to certain social welfare benefits in an attempt at limited income continuity.[28] The rationale for these additions may be obscure and fundamentally pragmatic but they illustrate the range of feasible compromises between an absolute flat rate idea and a full earnings relation which becomes available once the tort idea of maintaining the victim's standard of living is abandoned. For example, the nature and the level of income protection could now be designed to vary over time in a sophisticated way. It might be a better use of available resources to provide nothing or only a base level benefit for the initial period of disablement, increasing to a higher rate with full or partial earnings relation as longer commitments fall in. This most assists the long-term disabled, an espoused priority of reformers, which incidentally gives marginal preference to disease victims because traumatic injury leads to proportionately fewer

[27] Woodhouse Australia, para. 399.

[29] See J. Sleeman, *Resources for the Welfare State* (London, Longman, 1979), 102, 122–3; R. Titmuss, *Commitment to Welfare* (London, Allen & Unwin, 1968), 133–7; Pearson, paras. 111–13, 935; Atiyah, *ACL* 200 371, 397; Palmer, *Compensation for Incapacity*, 323–8.

cases of longer-term incapacity.[29] Alternatively, benefits with some degree of earnings relation might be provided in the short term decaying to a flat rate in the long term on the basis that this provides a short-term buffer period but that individuals should be encouraged to adjust to their new long-term circumstances.

But what of income loss as a focus for preferential benefits themselves? At the beginning of this section it was noted that before one could select a focus for preferential benefits the reasons for any preference for the physically disabled had to be clear. The problem for this author is the difficulty in advancing a cogent reason for such a preference at all, and, in particular, one that would lead to preferential benefits aimed at assuaging income disruption. Whether the separate disability benefits scheme was to be based on a policy of income support (flat rate) or income continuity (earnings-related benefits), it is hard to see why these benefits should be preferentially higher than those offered to those suffering income loss from other sorts of misfortune. Even if a rationale could be found, it would be a hard distinction to operate in practice, for example, when a middle-aged worker's employment is terminated because he is no longer physically strong enough for the requirements of the job or when priority in retrenchments falls on those partially disabled. The anomaly would be particularly stark in cases where workers suffer a loss of income because they have left a particular job to avoid a serious threat of disease but where there is no overt incapacity to attract compensation predicated on personal injury.

Option II: Disability as the Focus (see Table 7.1) In contrast to a personal injuries scheme which focuses on the misfortune of income loss, schemes may focus on the disability itself. In its simplest form this leads to a formula based on a flat sum, say £100, of which the beneficiary receives a percentage which reflects his percentage disablement.[30] The idea that disability *per se* is worthy

[29] Pearson, vol. 2 para. 108.

[30] Examples of such proposals include: H. Luntz, 'The Benefits Payable Under a Statutory Scheme', paper delivered to the *Accident Compensation: The Prospects for Reform* Conference held in Sydney, 4 Aug. 1983; Woodhouse Australia, paras. 390–400, using a flat rate incapacity-based formula for permanent partial disability; proponents of a uniform disability benefit cited by R. Lewis, 'No-Fault Compensation for Victims of Road Accidents: Can it be Justified?' (1981) 10 *J. Pol. Sci.* 161, 163 n. 7. See also The Disability Alliance, *Poverty and Disability* (London, Disability

of compensation regardless of actual impact on earnings or earning capacity—that a judge who loses a leg, a boilermaker who goes deaf should receive a benefit for the disability despite the lack of income loss—has a number of distinct practical advantages. Firstly, partial disabilities can be dealt with without the sort of discontinuities in benefits noted under flat rate income support schemes. Secondly, as a flat rate disability scheme does not involve an earnings base it does not involve the disincentive to rehabilitation and discrimination against non-earners that income-based systems do, while the abandonment of the earnings base simplifies the incorporation of deterrence goals into funding without the problem of regressive effects. But there is a major problem: where to draw the boundaries of 'disability' for the purposes of entitlement to benefits. Throughout the discussion so far this fundamental question has been glided over to facilitate the analysis but it should now be faced. How, in a comprehensive scheme which abandons a coverage rule based on cause, should a new coverage rule be constituted if, as implicit in a disability-based scheme, the impact on earnings or earning capacity is also abandoned as a criterion? One suggestion is that the new criterion should be one of functional loss, but this only reformulates the question—which functions are relevant for entitlement? For example, should the scheme cover sterility, speech impediments, dyslexia, diabetes, or incapacity to play a particular sport such as sky-diving because of proneness to dizziness? If so, would the relevant cases of partial functional incapacity be covered—is it intended, for example, to provide compensation for significantly low sperm counts? If, on the other hand, these cases are to be excluded, how can this be reflected in the coverage rule of the scheme?

Apart from the practical pros and cons, what is the merit of a disability focus? There is the apparent advantage that it is superficially easier to separate a disability-focused scheme from social welfare schemes handling other misfortunes with their income loss basis, and thereby to defuse pressure towards unplanned flow-on effects into these other areas (see 7.6). Indeed,

Alliance, 1975), 11. The most well-known example in operation is the flat rate industrial disablement benefit, although some earnings-related supplement, the special hardship allowance, was incorporated from the start: Atiyah, *ACL* 397; Pearson, para. 101, pp. 435–41.

proponents use the distinction to argue that the disability benefit should be independent of, and *in addition to*, other social welfare benefits such as unemployment benefit, pensions, etc.[31] But the argument is intrinsically elliptical because it still leaves us without a convincing rationale of why disability is itself a justified focus. If it is a concern for suffering *per se*, the idea runs into the same objections that damages for non-economic losses under tort attract: that is, that it is an idea at best unworkable and at worst an unjustified bias to non-economic suffering from one source. Why do these non-economic 'losses' of physical disability rate a higher social priority than those associated with other misfortunes?

On the other hand, disability benefits may in actual fact be an attempt to approximate income loss. Larson has convincingly argued that the origins of the earliest flat rate disability benefits (in industrial compensation schemes) were of this type: not the manifestation of a new principle of compensation for impairment *per se*, but an approximation to the central focus which remained income loss.[32] The approximation to income loss provided by a disability scheme based on a flat rate sum (Option II) is fundamentally crude: it is an approximation to the lost earning capacity of a standard man or woman; it would probably require the added complexity of adjustments like the present special hardship allowance under the industrial injuries scheme to cope with cases where actual income loss is significantly disproportionate to the degree of incapacity; and the approximation involves a marked bias to white-collar workers. On the other hand, it may be thought that this is a price worth paying for the overall advantages of such a simplified formula dispensing as it does with proof of pre- or post-disability earnings. But if economic loss is the true and sole basis of a disability benefit, we are then cast back to the dilemma of why income loss from this misfortune demands preference over other sources of income loss.

If the thesis of this book is accepted and the form of the tort remedy is no longer allowed to dominate the structure of benefits to be provided by a new public system of support for the disabled, then perhaps it would be more sensible, even if only on pragmatic grounds, to abandon strict notions of economic and non-economic losses as no longer necessary, particularly useful, or relevant. This

31 See e.g. Luntz, 'The Benefits Payable Under a Statutory Scheme', 18.
32 Larson, 'The Wage Loss Principle', 508.

merging of traditional categories would involve problems, for example, sorting out how to dovetail the compound benefits with social security benefits designed only to cover economic losses. But in the end it may be the most attractive approach if, say, preferential benefits for the disabled were regarded as adequately defended simply on the pragmatic basis of an inexplicably greater public concern for this misfortune rather than on some particular focus of the disabled's plight. If this is so, it is clear that disability-based schemes are well designed for such a broad brush compromise solution of an undifferentiated sum of support.

Option III: Combined Focus: Disability Benefit Geared to Earnings (see Table 7.1) The benefit formula could consist of two variables, the first being the victim's percentage disablement which is multiplied by a second factor reflecting the victim's pre-disablement earnings.[33] Because such a formula allows a higher earner a higher benefit it must, at least in part, be put forward as an attempt to approximate to loss of earnings/earning capacity to be at all defensible. Here, however, the approximation is far more inequitable and unpalatable than the formula in Option II. The formula has the drawbacks inherent in an earnings-based scheme —discrimination against, or complete exclusion of, non-earners, and potentially regressive effects (although, not being dependent on proof of actual loss of earnings, it does not inhibit rehabilitation) —plus the disadvantages of a disability focus such as the criteria of disablement. However, the most important flaw in such a formula is the compounded way in which it prejudices the less well off: it not only favours white-collar workers in the way its flat rate analogue Option II did, but it also favours higher earners. To illustrate this one must temporarily return to the notions of economic and non-economic losses. In these terms what an earnings-related disability-based formula attempts to do is to compensate both types of loss simultaneously. This is intrinsically inequitable. Suppose three right-handed victims suffer the same 30 per cent disablement by loss of their left hand. The benefit each will receive under Option III is 30 per cent of their pre-disablement earnings (see Table 7.2).

[33] A version of this formula now operates under the NZ scheme, 1982 Act (NZ) s. 60 (see Palmer, 'Accident Compensation in New Zealand: The First Two Years', 26), and in the Netherlands for earners (see Pearson, vol. 3, ch. 6).

Table 7.2 Examples of Benefit Payable Under Option III

Victim	Earnings Pre	Earnings Post	Actual earnings loss	Benefit
*V*1 (judge)	200	200	0	60
*V*2 (clerk)	100	100	0	30
*V*3 (labourer)	100	50	50	30

In cases where there is no income loss (*V*1 and *V*2) the benefit, which can only be seen as assuaging non-economic losses, is unjustifiably earnings-related and favours the higher earner. Similarly, comparing *V*2 and *V*3, the two victims receive the same benefit for the same injury yet the white-collar worker *V*2 has in fact suffered no income loss. It follows that *V*2 receives a benefit of £30 for his non-economic losses while *V*3 receives nothing for his non-economic losses and the £30 he does receive does not even cover his actual loss of income.

Under Option II I noted that, once tort is abandoned as a guideline for benefits, it may make sense to reject an analysis in terms of economic/non-economic losses and to adopt the compromise of a disability-based global sum to assuage the undifferentiated impact of disability. But once an earnings-related element is injected into the formula, it inescapably reintroduces notions of economic loss. It is then far more tenable to see the benefit as an approximation to economic loss than a global sum for the disabled, and, that being the case, the sort of biases involved are distasteful, particularly as they favour groups of earners who are already those better served by the economic system. Indeed, it was to avoid these compounded biases that Woodhouse, himself a strong advocate of the earnings-related principle, abandoned it in cases of partial disability, preferring a flat rate disability base.

Option IV: Two-Tier Benefits One way around these biases is to use a two-tier system of benefits, one benefit being aimed at loss of earnings/earning capacity (either flat rate or earnings-related), and a separate benefit aimed at the non-economic aspects of disability, that is, a percentage disability benefit like that described under Option II.[34] It follows from the foregoing discussion that this latter

[34] For examples of two-tier systems or proposals where the income benefit is flat

disability benefit could only justifiably be flat rate. Thus a boilermaker who goes deaf may receive nothing under the first benefit but still receive some support under the second benefit scheme to assuage the loss of amenity, etc., brought about by his disablement.

Another, and completely different, form of the two-tier device operates in the Netherlands for long-term disability. Here the general social security system provides citizens with an invalidity benefit based on a flat rate disability formula (Option II), but tops this up in the case of earners with an earnings-related disability formula (Option III).[35] There is no separate benefit covering work-related disability.

Option V: The Time Periods Approach One particularly useful device in selecting what benefits to institute is to have a different focus of benefit according to the length of time the person has been disabled. One variant of this was discussed under Option I—the idea that whether a benefit for income loss was to be flat rate or earnings-related could vary according to the length of time the recipient had been disabled. More interestingly one could adopt, say, a generous income focus for the first six months—thereby minimizing the pragmatic problems in the contraction of rights currently enjoyed by certain earners—and thereafter a flat rate disability focus for the few who remained disabled. Advantages of the approach are that it allows an adjustment period for victims, limits the drawbacks of an earnings-related scheme to the first six months, and is consistent with the general view that benefits for disability *per se* are best applicable to, and most justified in, cases of long-term disability.[36] However, the idea that a disability benefit is more appropriately reserved for the long-term disabled, because it is only after a significant time that disability is adequately stabilized for assessment purposes, does not take into account those cases—which would be numerous in any scheme which covered disease—where the condition is intrinsically fluctuating or

rate, see Disability Alliance, *Poverty and Disability*, or earnings-related, see Ison, 'Human Disability and Personal Income', 432; Gair committee, paras. 80–1.

[35] Pearson, vol. 3, pp. 118–19. For criticism see 3.7 and 5.8.

[36] Luntz, 'The Benefits Payable Under a Statutory Scheme'; Harris *et al.*, *Compensation and Support for Illness and Injury* (Oxford, Clarendon Press, 1984), 337, who suggest three time periods.

erratic in its degree of disablement over time. Considerable reformulation of basic definitions would be necessary to give effect to such a scheme (see 7.8).

7.8 *A Suggested Approach: The Non-Comprehensive Option*

I have argued that the abandonment of a causal criterion in the treatment of the disabled fatally undermines the justification for using tort as a guide to the appropriate form and content of a benefits scheme. In the last section a broad range of options for a comprehensive scheme was reviewed. This was done at some length, not simply because it is worthwhile to bring together in one place major options and their pros and cons, but to illustrate the variety of approaches which could be adopted under a unified comprehensive system. But there is a further step to be taken in freeing the analysis from the influence of the expectations generated by the tort remedy, and that is to question the assumption that tort should be replaced by a separate comprehensive scheme—that is, a scheme covering *all* cases of disablement, long-term and short-term, partial and total. Once the causal criterion, and hence the tort influence, is abandoned, this basic tenet of modern reformers is now open for reassessment.

The comprehensive goal may be postponed—as it would be, for example, in the option discussed earlier (7.6(*a*)) where a first step of disease coverage was suggested among other things as a device to defuse the resistance of vested interests. The idea of comprehensive cover may itself be rejected outright in the light of practical and/or policy considerations. For example, an opting-in scheme somewhat along the lines suggested by O'Connell (see 5.4) might be used. The public fund could offer a certain level of benefit in return for some statutorily sanctioned assignment of any possible claim in tort held by the victim arising out of his or her disablement. In turn, the fund could offer alleged tortfeasors a settlement at a level considerably lower than the likely common law damages in the case. Such a plan has distinct advantages. For example, in addition to those described by O'Connell, the scheme would have the dramatic advantage in Australia of avoiding the constitutional flaws implicit in any federal scheme which required the compulsory abrogation of common law rights, for here there is no compulsion on the victim to opt in and thereby lose his common law rights.[37] But the drawbacks remain: the scheme relies on the engine of tort liability so that it leaves the wasteful common law/liability insurance system intact and the financial basis of the scheme would probably be severely undermined by the process of 'adverse selection', that is, the only people who would be likely to opt in are those with a poor or no chance of getting the higher benefits under the tort/settlement process.

A third and more dramatic departure from the conventional comprehensive doctrine adhered to by reformers would be an approach which involved not one fully integrated system of comprehensive coverage and benefits—but co-ordination of separate schemes for specific needs, for example, medical costs, income protection, out-of-pocket expenses, retraining expenses, even non-economic losses.[38] The advantages of this option are considerable. The separation of needs allows greater flexibility in their treatment, reflecting the different policy factors applying to different types of need. It would also seem possible to achieve more coherent integration with other forms of social security geared to heads of need (misfortune) such as income loss. For example, the income protection fund might take responsibility for a certain percentage of that need (a percentage independent of the percentage protection deemed appropriate in other contexts like unemployment). This percentage could vary over time and a different percentage be allocated to another source such as an employer who might be required to pay a mandatory benefit for a certain period after which fund benefits took over. If a coherent rationale for distinguishing between causes of wage loss could be found, such a system would also seem better able to structure benefits in a way which revealed, and put into effect, such policy differences between, say, sick-pay, redundancy, and unemployment pay. The separation of needs would allow more emphasis to preferred subclasses of the disabled, such as those totally incapacitated and/or those with long-term disability, according to the particular scheme. One scheme, such as the income protection fund, might concentrate on a particular subclass, say the totally disabled, while others, such as the scheme directed at retraining,

[37] Presumably this is also the constitutional basis of s. 135R of the Social Security Act (Cth.) 1947. On constitutional flaws of the Woodhouse plan see Senate Report, para. 3.4.

[38] This broad option was noted as a possibility by Professor Atiyah.

could be available to a different, wider subclass, say the very long-term disabled.

By looking a little more closely at one rough model of this third non-comprehensive option, we will see how policy choices which are not in themselves remarkable can produce an end-result far from the currently accepted wisdom that the optimal solution of the disability problem is a comprehensive scheme. It will also be instructive to see at the same time how giving the issues raised by cases of non-traumatically caused disability the attention their numerical dominance deserves can place specific demands on, and be reflected in, the model. We start with the set of current assumptions: that a distinction based on the cause of disability is unwarranted and that a coherent rationale is available for giving the disabled both separate and preferential treatment from the victims of other misfortunes.

(*a*) *The Income Protection Scheme* It is arguable that the class of disabled in greatest need are those who are totally or severely handicapped and that the most serious aspect of this misfortune is incapacity to earn. A non-comprehensive system could focus on this by having its income protection scheme restricted to cases of total incapacity for work, providing no benefit for disability *per se* or for partial loss of earnings. Cases of severe disability which yet do not produce total loss of earning capacity—the judge who loses a leg, the boilermaker who goes deaf—would still be entitled to assistance under the other schemes for separate needs such as retraining, but priority for income protection could be restricted to the totally incapacitated, at least in the initial phase. There would be problems: should an income protection scheme attempt any earnings relation or not; should its level of assistance vary according to the length of total disablement; how would it blend with associated income support systems for other misfortunes such as pensions and unemployment benefit in such a way as not to deter rehabilitation, etc? On the other hand, this model does seem well suited to the priorities suggested by non-traumatic disability. From their national study of household disability in Britain, the Centre for Socio-Legal Studies in Oxford concluded that, *inter alia*:

(1) Illnesses lead to far higher losses of working time; and

(2) Those most prone to labour market disadvantage following disability are those who have suffered prolonged absence from work irrespective of the extent of residual disability, so that benefits are more appropriately based on absence from work *per se* than on measured health impairment.[39]

Giving priority to the problem of total incapacity for work would, therefore, take good account of these characteristics. It also simplifies a number of technical problems in providing benefits to victims of non-traumatically caused disability. The principal difficulty here stems from the different pattern of disablement experienced by victims of non-traumatic injury. In an accident the victim typically suffers suddenly the maximum disablement he will experience in the course of the injury, a disablement decaying away to a complete recovery or to a relatively stable level of permanent incapacity. In contrast, the disablement of a disease victim usually increases gradually and often does not reach a stabilized condition. The disease is either progressive and degenerative or, like asthma and epilepsy, recurring intermittently. This characteristic of fluctuating disablement creates great problems for assessment in both earnings-related or purely incapacity-based schemes. How is either to handle cases where the percentage disablement and loss of earning capacity ranges from 100 per cent through to full remission and back again—particularly given the common uncertainty of disease prognoses? For example, while in an accident scheme it makes sense to have a provision allowing reassessment of incapacity in only one direction (namely, upwards, to avoid inhibiting rehabilitation), such a provision would be unduly generous in a scheme encompassing large numbers of fluctuating disabilities. The drastic solution to the problem of fluctuating disability suggested by Woodhouse (Australia) was to exclude benefits for permanent partial disability *in disease cases*. But this creates a distasteful anomaly with the treatment of accident victims, an anomaly which is in no way rationalized by Woodhouse's boldly stated justification that 'it would be wrong to make regular payment to persons with chronic problems such as diabetes' except during periods of total incapacity.[40] There would also be the

[39] Harris *et al.*, *Compensation and Support for Illness and Injury* 263 and 277–8.

[40] Woodhouse Australia, para. 400 (i).

practical problems and anomalies associated with the accident/process distinction which would necessarily be reintroduced by such a rule. The more sensible solution outlined in the model non-comprehensive system would be to abandon the arguably attractive goal of compensating partial disability for the sake of ensuring the adequate and workable cover for *all* cases of total incapacity. This has the advantage that the problem of fluctuating disability and uncertain disease prognosis would no longer be relevant to entitlement which would simply depend on total incapacity for work (although it would still influence the question of the required frequency of the periodic reassessment of disability which is necessary to inhibit fraud and malingering).

An associated problem with the pattern of many cases of non-traumatically caused disability is that of qualification periods. Take the distinction routinely made between short- and long-term disability. It is an espoused priority of most reformers to give the long-term disabled higher priority. The definition of short- and long-term disablement is assumed to be expressible in terms of length of continuous incapacity, say six months total incapacity for work. The preference for the long-term disabled is then reflected in benefits (or increments to existing benefits) which do not become available to a disabled person until this qualification period is completed. Say, under the suggested income protection scheme covering all cases of total incapacity, this was to be reflected in an increase in benefit after six months continuous entitlement. The victim of a fluctuating condition or a disease like cancer, which is liable to periods of remission, may often be rendered totally incapacitated—and therefore qualify for the short-term benefit rate—but never establish an unbroken record of total incapacity for six months to qualify for the benefits designed for the long-term disabled, even though, of course, they may well be as disadvantaged by their condition—say in securing and retaining employment—as a victim whose incapacity is total and permanently stabilized at that level. Reformers must reassess their use of the length of continuous incapacity as a reflection of the gravity of the misfortune in the individual case. The proposed system of benefits would, therefore, incorporate entitlement provisions sufficiently flexible to accommodate these cases—perhaps by allowing a suitably qualified scheme of accumulating separate periods of related incapacity.

(b) Schemes Focused on Rehabilitation, Retraining, and Other Needs To back up the income protection scheme there would be separate systems providing relevant benefits and services to meet other needs of the disabled. These systems could be open to all disabled, including those only partially disabled, and regardless of whether the disability affected earning capacity. Some of these systems would be focused on needs peculiar to the disabled, such as the need for health care in general, physiotherapy, medical aids, guide-dogs, reading aids, special diets, special housing costs (for example, alterations to residences or vehicles), and home nursing care. Other needs of the disabled are also shared by victims of other misfortunes: for example, both the deserted parent and the disabled parent may be in need of child care facilities and home help; both the worker made redundant and the injured person may be in need of rehabilitation grants or, say, low interest loans to enable retraining. Separate but co-ordinated systems could be designed to meet the individual class of need and structured in such a way as to reflect any scale of priorities between classes of claimant. For example, there may be a system providing benefits to assist with certain costs of living such as rent, telephone, and travel costs and the assistance rendered to the disabled claimant may be at a higher level than other classes of disadvantaged people. A telephone is of greater importance to a blind, aged person than to a young, healthy, and mobile claimant. On the other hand, the priorities may be reversed in other cases: the long-term unemployed would seem to have more claim on retraining resources than a case of short-term partial disability. The more generous these back-up services for the disabled and the more they are geared to long-term disablement, the better the overall system will serve the dominant subclass of the disabled—the victims of non-traumatically caused disability. This is supported by two more conclusions by the Oxford survey team:

(1) Victims of illness are more likely to suffer residual incapacity than accident victims; and
(2) On the basis of use of medical and health services the 'needs' of illness victims appear to be greater than those of accident victims.[41]

[41] Harris *et al.*, *Compensation and Support for Illness and Injury*, 242 and 323.

(*c*) *Funding* The system could be funded by both general revenue and contributions from activities identified as producing disablement such as motoring, smoking, etc. Benefits would not be predicated on such identification in the individual case, but the administration of benefits would produce a central register of disablements, perhaps linked to a centralized system of general medical records, which could be designed to facilitate identification and therefore deterrence of man-made causes of disability by the central pooling of data. In short, the system would both pool existing data and generate more sophisticated data. The degree of deterrence feasible is intrinsically limited (see Chapter 6), but such a system would help to optimize the amount of deterrence achieved per pound or dollar invested by government in the use of mechanisms which deliver benefits and services to achieve deterrence. The more successful the system is in attracting claimants, the better the available data base and the better the prospects for deterrence, both by contribution rates by injurers to the fund and by direct regulation. The more attractive the level of benefits and services, then, the more likely people will be to claim and thereby have this effect (see 7.9). As to the level of contributions, added deterrence can be effected by charging more in contributions than is paid out by way of benefits for that source of disablement. This could be used to deter the full range of disablements caused by that cause even though only total incapacity for work qualified for income protection benefit.

Finally, there are two points of difficulty in financing a system which extends to disease cases. The first, which is not completely specific to disease victims, is that if the income protection benefit is to be earnings-related and funded on an insurance basis from contributions siphoned off from payroll or earnings, then the congenitally or chronically disabled and those suffering from fluctuating disabilities may find it impossible to build up eligibility. There is no easy answer to this. Secondly, inclusion of cases of non-traumatically caused disability raises the question of how well the extent and distribution over time of future claims arising from latent disease (see 6.5) could be predicted, and hence how feasible it would be to finance the income protection scheme on a fully funded basis. Are current morbidity statistics sufficient for this purpose?

(*d*) *Lessons to be Drawn from Sketching such a Proposal* It can be seen from this brief outline of a possible non-comprehensive system that abandonment of tort as a guide to reform allows wide flexibility in the choice of priorities. It shows that by taking a few assumptions, unremarkable in a welfare context—for example, that, at least initially, most support should go to the severely incapacitated even at the price of excluding another subclass such as the partially disabled—a reform proposal is produced which is both far different from those which have emerged from the 'compensation debate' and better focused on the needs of the majority of the disabled, the non-traumatically disabled. That looking at the disability problem from such a purely 'welfare' perspective produces these results and advantages, is confirmed by the form of the current provision under the social welfare system for non-industrial disability, which is broadly similar to that of the non-comprehensive proposal (flat rate benefits predicated on total incapacity to work plus benefits for special needs such as attendance allowances, mobility allowances, etc.).

This emphasizes the importance of ridding the disability debate of the lingering shadow of the tort remedy, as well as suggesting that the current form of welfare provision provides a healthy basis for future improvements in the support of the disabled. To synthesize an elegant, workable blueprint for such development will be a mammoth task and is well beyond the scope of this book. It will, for example, have to address complex questions such as how to abolish the industrial preference, and will have to come to terms with the fluctuating popularity of the earnings-related principle. What is more, the process will have to take place, if at all, in an era devoid of the sort of political consensus prevailing in the years when Beveridge and Woodhouse considered the questions —the turmoil of the current political dynamics of the welfare state being well illustrated by the 1985 Green Paper on the Reform of Social Security and the subsequent attacks on its miminalist approach.[42]

[42] *Reform of Social Security* (June 1985) Cmnd. 9517; D. Donnison, 'The Backlash to Beveridge', *Observer*, 9 June 1985; Child Poverty Action Group, *Burying Beveridge* (London, CPAG, 1985).

7.9 *The Fate of Existing Entrenched Rights: A Separate and Peripheral Question*

We have looked at the distortions of the narrow 'compensation debate' and the proposals which have emerged from it. Rejecting these, we moved on to look at the question of what could be done for the disabled starting from a welfare perspective, and some options for the basis of reform were reviewed. Before concluding by looking at whether separate 'disability reform' is justified within the wider welfare context (7.10), we must tie up one loose end. Tort lawyers may be distressed by the unanswered question of what to do about existing entrenched rights, that is: what to do with those superior rights enjoyed now by a minority—recipients of tort damages and industrial compensation—which developed and became entrenched at a time when the welfare state provided little or no support to the disabled. No satisfactory answer can be given to this question until one appreciates the extent to which the privileged position of these groups is a result of broad ideas about welfare and compensation rather than the rights of individual victims.

In the early twentieth century developing ideas of society's role in compensating victims of personal injuries were reflected in the establishment and expansion of no-fault industrial compensation and in moves to improve tort as a compensatory mechanism by, for example, imposing statutory obligations to insure against tort liability and by abolishing the industrial defences. When these welfaristic ideas were eventually given comprehensive public consideration and expression in an independent fully fledged welfare system, the level of support settled on as appropriate for the non-industrially disabled was dramatically lower than the level of compensation provided by tort and the new version of industrial compensation. So it can be seen that, ironically, the privileged position of tort victims and the industrially disabled is a manifestation, at least in part, of essentially the same ideas which, when developed rigorously in terms of comprehensive social policy and priorities, lead to quite a different result: to benefits for a much larger group of the disabled but at considerably lower levels.

Reformers took on board these welfare ideas to a degree sufficient to find the anomalies between the privileged disabled and the rest distasteful (see 5.9). But, as I have argued, the logic of these needs-based attacks on tort does not necessarily lead to the replacement of tort by a comprehensive Woodhouse-type scheme, which is the current wisdom. What a needs-based approach to disability does lead to is the broad question: *What should a welfare state do for the disabled?* The considered answer to this may involve, for example, a non-comprehensive solution such as the Beveridge approach to sickness benefits, or the similar option outlined above, under which some of the disabled who currently receive benefits through the tort or industrial compensation system (for example, the partially disabled) would not be provided for at all, on the basis that there are other groups (for example, the totally disabled) who have stronger claims on limited resources. The point here is that whatever answer one gives to this crucial question, it should not depend on one's attitude to the future of tort or the industrial preference. It is true that abolition of such a wasteful remedy as tort and the anachronistic industrial preference would free considerable resources for more socially equitable purposes. But the fact that tort is a wasteful and inefficient system or the fact that the rationale for the industrial preference is unclear, while providing good reasons to reform or even abolish them, are issues quite separate from, and of far more limited importance than, the basic welfare question of what the state ought to do to meet the needs of the disabled.

7.10 *Ramifications for the Social Policy Debate: Is Preferential Disability Reform Justified?*

The reassessment of the debate on compensation for personal injuries from the viewpoint of disease victims has finally led to this most fundamental issue: what should the modern welfare state do about the disabled? The first question relevant to this issue is: should a distinction be made within the class of disabled on the basis of the cause of disability? Virtually all commentators and reformers reject such a distinction on principle and this is certainly the conclusion to which the needs-based logic of reform leads. It must be acknowledged, however, that the persistence of the accident preference in reform proposals may reflect a pragmatic preference for short-term and limited rationalization of resources over grand comprehensive schemes (see 7.6(*a*)). Nor can such an *ad hoc* compromise approach be dismissed simply as irrational and indefensible—it may simply be the best that is feasible given more pressing social concerns. Indeed, to some, short-term pragmatism may itself be a more desirable and attractive way to run society; for example, it may be thought that short-term planning is more responsive to the democratic process. On the other hand, it is important at least to know what conclusions certain espoused principles would lead to if worked out logically and fully. On that basis it can be assumed here that, as no justification on principle can be found for a distinction in treatment within the class of the disabled on the basis of cause of disability, none would be made in any reformed system of provision for the disabled.

The second question which must be answered is this: once public assistance to the disabled is no longer seen as linked to the remedy in tort, should the disabled be treated by the state differently from victims of other misfortunes? As noted earlier (7.7), this is a major threshold question, because its answer will not only determine whether there is any justification for a separate scheme of benefits for the disabled, but form the basis of the selection of the form, content, and coverage of any such scheme from the broad range of options theoretically available. Certainly many reformers see reform in the provision for the disabled as a major social priority, but to get such reform into context one must ask: is there any relevant difference between physical disablement and other misfortunes such as poverty, unemployment, or income loss suffered on retirement or desertion by a supporting spouse, which are all misfortunes for which state benefits are available? Is there a rational basis for the current unexplained differential between treatment of disablement and other misfortunes both in the compensation debate and within social welfare systems generally?

Certain arguments stressed by reformers in support of better compensation for accident victims might be called in aid to defend this preference for the physically disabled. One such argument is based on the notion of 'community responsibility', but just as this does not, for example, dictate that a public system of compensation should be *funded* on a flat rate basis as Woodhouse seemed to conclude, so too it does not alone justify preferential benefits to the disabled: 'community responsibility' may indicate a duty to act but not the method or the level of protection which ought to be provided by the state against a particular class of misfortune. Nor do simple appeals to the rather vague notion that the toll of accidental personal injury (or disablement in general) is a great and pressing social problem[43] explain its preferential treatment over more common misfortunes of greater economic impact on the individual such as unemployment. Similarly, despite the emphasis of reformers, the fact that disablement can produce a sudden interruption to the victim's standard of living does not differentiate that misfortune from others such as unemployment or, say, the desertion of a wife with dependent children. Finally, references to the fact that certain personal injury is inevitable or an incident of complex modern social organization do not justify preferential treatment of the disabled over victims of other equally inevitable 'diswelfares' or 'disservices' of a rapidly changing society such as the obsolescence of acquired skills ('deskilling'), unemployment, and urban blight.[44]

In terms of abstract principle, it seems likely that, as Blum and Kalven have suggested, at the root of these arguments about when the state should do something about human misfortune is a concern with poverty.[45] Certainly it appears that the basic arguments used by lobbyists when pushing for better provision for the disabled concern themselves with the victim's finances. In arguing for a comprehensive income scheme for the disabled the Disability Alliance, for example, cites three characteristic features of the disablement victim's financial position which call for a special approach.[46] First, the income of a large number of the disabled falls below the government's official poverty line, a line reflecting minimum needs. Secondly, the finances of the disabled are often depleted by special needs and expenses which are attributable to the disability. This means, says the Disability Alliance, that 'many of the disabled have a need for *higher* incomes than the non-disabled'. Thirdly, there is the problem of the current inequalities of treatment by the state in providing income to different categories of the disabled; for instance, the congenitally disabled are provided for at a much more modest level than, say, disabled earners, because they do not qualify for

[43] Woodhouse NZ, paras. 2, 17; Commentary, para. 1; Woodhouse Australia, paras. 2, 52–3, 427. Compare T. Ison in 'The Politics of Reform in Personal Injury Compensation' (1977) 27 *Univ. of Toronto L. J.* 385, 398, arguing for the urgency of the plight of disabled in general, with attitude of Royal Commission on *Social Security in NZ*, referred to earlier in n. 11.

[44] Titmuss, *Commitment to Welfare*, 63, 133.

[45] W. Blum and H. Kalven, 'Public Law Perspectives on a Private Law Problem: Auto Compensation Plans' (1964) 31 *Univ. of Chi. L. R.* 641, 721.

[46] Disability Alliance, *Poverty and Disability*, 7–11.

benefits under a contributory scheme. In my view, none of these arguments justifies separate treatment of the disabled. The third problem could be significantly resolved by abolishing differential treatment on the basis of cause, and limiting anomalies created by the generosity of contributory benefits either by abolishing contributory schemes or improving the benefits available under non-contributory schemes. Secondly, and ironically, the argument from anomaly defeats the first argument based on the fact that many of the disabled fall below the poverty line. Poverty is not peculiar to the disabled—just because disablement is a particular source of poverty does not justify separate and more generous treatment of those people whose poverty is attributable to this source. Finally, the second problem calls for an adequate system of special grants and services to cover these extra needs but does not justify preferential access for the disabled to programmes, such as retraining schemes, to which the victims of other misfortunes such as the deskilled also need access.

Even if it were accepted that the root of our concern with the disabled is their financial plight, it would still be necessary to decide who belonged to the class of disabled persons for the purpose of giving them preferential treatment. In practice, this will create problems in cases where a person is a victim of more than one misfortune. Take, for example, the case of a partially disabled person whose loss of a job or inability to find work is a combined result of his disability and the state of the job market. Similarly, if a person 'retires' early because he or she is not physically up to the requirements of the job, is the income loss due to disablement or retirement? Not only do such borderline cases produce difficulties in practice but they also raise once again in acute form the question of the social justification for according preferential treatment to the misfortune of disablement. As Liebman, who queries the preferential treatment of physical disablement within the United States social welfare system, says:

> Our society may revere medicine and symphathize with the sick, but it holds no views that could explain distinctions between persons totally unable to work according to whether their condition results from an illness or, on the other hand, from limited natural abilities, decades of racism or sexism, homosexuality, family burdens, technological change, a broken home or national fiscal policy.[47]

[47] L. Liebman, 'The Definition of Disability in Social Security and Supplemental

In the end, there seems to be no reason intrinsic to the misfortune of physical disablement which would justify its being given special treatment by the state over other misfortunes. Nevertheless, we must now ask, even if this is so, are there other reasons for treating disablement separately and/or preferentially? There seem to be two such possible extrinsic reasons. First, it may be worthwhile from a deterrence point of view.[48] Unlike many other misfortunes in which the state takes an interest, in some cases of disablement there is potential for deterrence. By having a separate scheme for disablement benefits, data could be collected and generated improving the chances of accurately identifying the source of the man-made risks (for example, a particular product or aspect of life-style), sheeting home to that source the appropriate cost, and, in some cases, directly reducing the risk by government regulation or other techniques (such as a general health policy on nutrition, provision and distribution of health care facilities, etc.) (6.8). Benefits for disablement could be set at a sufficiently generous level to attract claims which might otherwise be made for the general benefits available for income support. The separate scheme would then be justified in terms of data collection for deterrence/prevention purposes and the preferential benefits would be justified in terms of providing that process with the necessary data of claims.

A second extrinsic reason which may explain (if not justify) a social preference for the disabled is the fact that, when making what Calabresi and Bobbitt term 'tragic choices' about the allocation of scarce resources between fundamental and highly desired ends, people sometimes act on the basis of irrational sympathies or psychological preferences rather than reasoned argument.[49] For example, while it might seem more sensible to spend £1 million on the maintenance of mines than on saving a single miner trapped by a cave-in, our sympathy for the individual in distress makes us willing to go to very great lengths to save him. The fact that the price of our concern is that many other miners

Security Income: Drawing the Bounds of Social Welfare Estates' (1976) 89 *Harv. L. R.* 833, 864.

[48] Compare footnote 10 above: Calabresi arguing that the accident preference may be justified in terms of deterrability.

[49] G. Calabresi and P. Bobbitt, *Tragic Choices* (New York, Norton & Co., 1978).

may have to work in riskier conditions does not seem to throw any practical doubts on the decision to help the individuals. Indeed society's conscience would be outraged by any other decision.

Similar attitudes may be at work in relation to compensating the physically disabled. Although no rational argument seems to justify a preference for the physically disabled, it may be that the nature of their disabilities and the source of their misfortune excites sympathy in a way that the plight of, for example, the mentally disabled or the unemployed does not. Even if there is validity in this 'gut reaction' *explanation* for a preference for the disabled and other particular social choices and policies, it remains legitimate to ask to what extent basic issues of human welfare can and ought to be decided according to such pragmatic juggling of resources without regard to reasoned long-term goals.

Conclusion

The neglected issues raised by cases of non-traumatically caused disability are not peripheral to the current compensation debate. Nor does their examination lead to a simplification of that debate—indeed, the reverse is true. If account is taken of these cases, their numerical dominance inevitably shifts the debate from the current distortions of an accident-focused 'compensation' debate to a broad social welfare context. This necessitates a fundamental reassessment of available reform options and ultimate goals, for example, whether the prevailing wisdom of reformers in favour of an integrated comprehensive scheme for personal injuries based on a generous level of benefits is the goal to which logic and policy need inevitably to be pointing. Until there is a re-evaluation of such fundamental issues as why, if at all, the disabled should be treated preferentially over victims of other misfortunes, there will not be much gained from formulating detailed designs for schemes of benefits. The daunting lesson to be learned from a disease focus is that the 'compensation' debate is fundamentally and disturbingly more complex than we have generally assumed.

Index

The manufacturer's authorised representative in the EU for product safety is Oxford University Press España S.A. of el Parque Empresarial San Fernando de Henares, Avenida de Castilla, 2 – 28830 Madrid (www.oup.es/en or product.safety@oup.com). OUP España S.A. also acts as importer into Spain of products made by the manufacturer.

www.ingramcontent.com/pod-product-compliance
Lightning Source LLC
LaVergne TN
LVHW101441200926
848483LV00001B/15
9780198255529